St
in

STARTING STATISTICS
in Psychology and Education

A STUDENT HANDBOOK

S. Heyes, M. Hardy, P. Humphreys and P. Rookes

WEIDENFELD & NICOLSON · LONDON

Reprinted 1988
Reprinted 1990

Weidenfeld and Nicolson Ltd
91 Clapham High Street, London SW4 7TA

ISBN 0 297 78843 4 paperback

Phototypeset by Deltatype Ltd, Ellesmere Port, Cheshire
Printed in Great Britain by
Butler & Tanner Ltd, Frome and London

Contents

Preface

Many students of Psychology and other Social Sciences approach statistics with a certain amount of trepidation. Some of them chose their courses without realising that they would have to deal with numbers. Statistics are used as a tool to help describe and analyse the results of practical work. We have found that once students start to do the same and realise that they don't need to derive formulae or do much more than follow a few simple 'recipes', the fear of statistics disappears and they develop an intuitive feeling for what the statistics are doing, without any need for formal mathematics theory.

The important thing is that you learn to recognise when to use statistical tools and to interpret results obtained from them in terms of your research. If this means that the actual calculation of the tests is done by a friendly mathematician or a computer, then fair enough; nobody is ever going to ask you to remember formulae, and the cookbook-style 'recipes' for the tests will mean that most people will be able to do the calculations following each step with their calculators.

This handbook is meant to be used as an addition to formal lessons rather than as a replacement, though we believe that it can stand on its own for those who want to use it that way.

Of course there is no point in learning about statistics if you can't design a decent experiment or other kind of study to produce the data for analysis. Research design is largely a matter of some specific knowledge combined with a lot of common sense, but if you get this stage wrong no amount of juggling with statistics is really going to sort out the problem. This book is designed to give you the specific knowledge needed to organise your common sense so that you can carry out successful practicals and write them up in a way that others will understand.

We would like to thank Eileen Heyes, Tracy Rookes and Tricia

Heyes for their help in getting this handbook together. Thanks also to Jo Freeman and Pauline Simm who typed some of the original handouts from which the handbook was developed.

An overview of statistics and their uses

'There are three kinds of lies: lies, damned lies and statistics.' (Disraeli)

'You can prove anything with statistics.' (Anon)

How many times have you heard statements like these? There is a grain of truth in them – it is possible, sometimes, to pull the wool over people's eyes by presenting your statistics so that they put your case in the best possible way.

But it is actually quite reasonable to present statistical information in the way that backs up your argument best, especially when you include all the data that allows the reader who does more than

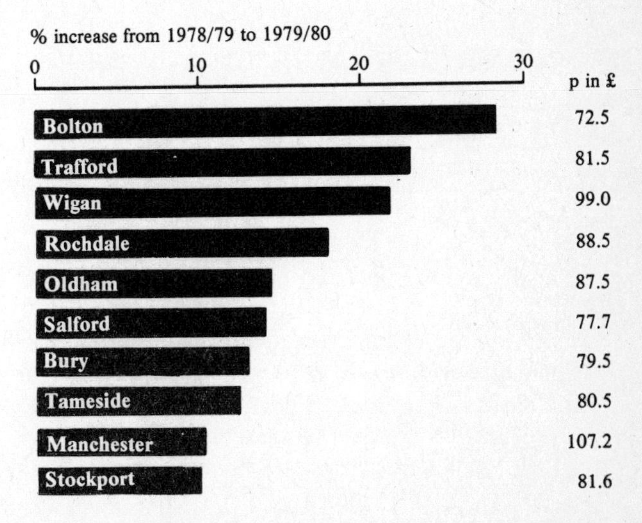

Figure 1. Domestic rates, Great Manchester Council authorities (*Source:* Metropolitan Borough of Stockport, *Civic Review*, April 1979.)

skim the page to get a full picture. As an example of the different ways that the same information can be portrayed, depending on what impression you want to make, look at Figure 1 which shows a chart published in *Civic Review*, the Stockport quarterly publication for ratepayers, in April 1979.

The chart is a truthful representation of how well Stockport had done in keeping down the rate increase; it also shows (down the right-hand side) that ratepayers in Authorities like Bolton and Salford paid less rates in the pound than did Stockport ratepayers.

Imagine, though, that you wanted to present the same information but this time wanted to put Bolton Council in a good light. Perhaps you would publish the chart reproduced in Figure 2.

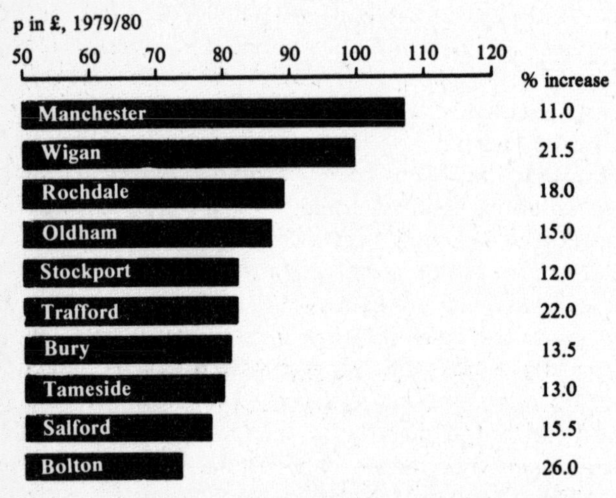

Figure 2. Domestic rates, Greater Manchester Council authorities (*Source:* As for Figure 1.)

The two charts present the same statistical information but the immediate impact is different because this time the bars on the chart have been used to denote the actual rate in the pound rather than the percentage increase. As readers, we need to study more than the visual impact of such presentations.

Some people appear to shut their eyes or turn off all critical faculties when presented with numerical information, yet these same people may show high intellectual abilities with other types of argument. It is possible to 'prove' anything to a statistical halfwit, but most of us, with very little effort and information, can become just as good at understanding and spotting the flaws in a statistical argument as we are when reading a purely verbal report. This is not surprising because we have been using statistics since the moment we were able to count.

Statistics are about describing the world in terms of numbers and making evaluations and predictions based on those descriptions.

Most people use statistics without even realising it – when you go out to buy clothes and give your measurements, when you work out how many miles to the gallon your car does, when you work out whether you have enough money to catch the bus *and* have an extra drink. Young children love statistics – counting the number of legs on a dog or the number of times they can bounce a ball, arguing that they are four and three-quarters rather than four, deciding who is tallest or who has got most marbles. At this stage there is no fear of numbers, although it is quite easy to fool them as those who have come across Piaget's theories know. For example, a child of five, when presented with a choice between a short row and a long row of chocolate buttons each containing five sweets, will usually choose the longer row, believing that this gives him most to eat.

There are two main types of statistic: **descriptive statistics**, of which you have already had a great deal of experience, both in school and in everyday life – though you may not have thought of them as statistics – and **inferential statistics**, of which you've probably had less experience. However if you've ever laid a bet on the outcome of some sporting or other occasion on the basis of your previous experience of the competitors, you've probably been considering an informal type of inferential statistic.

Descriptive statistics

Whenever you measure, weigh, time or count something, you are dealing with descriptive statistics; you are describing part of the world in terms of numbers. You have been doing this since the age of two. Descriptive statistics help us to understand the world as it is. When presented well, they reduce the complexity of the world, and can be used to give short summaries which allow a reader to quickly grasp the essence of a situation. We've all come across the idea of

average: how much quicker it is to be told that a batsman's average this season is 30 runs per match than to be given every individual score. This is especially so when we want to compare two or three people's scores.

You probably learnt how to work out an average when you were in primary school; you can probably still remember how to do it. Work out the average of the following numbers:

2, 2, 3, 5, 8

You probably worked out the answer to be 4 by adding up all the numbers and dividing by the number of numbers. This average is known as the **mean** average. But did you realise that the average of the above numbers is also 2, or could even be 3? This is really showing that statistics can prove anything! The problem is that most of us learnt what a mean average is at an early age but have not been told that there are other forms of average, such as the **mode** (the most common number in a list) and the **median** (the number that is exactly in the middle when all the numbers are placed in order of size). The concepts of mean, median and mode are dealt with in detail on pages 11–12.

All three forms of average are equally valid but in many cases they do have different values, and so you should always ask which sort of average someone is using – they may simply have chosen the one they have because it supports their argument best. You may have wondered why the reported average wage of some workers on strike seems to vary depending on which newspaper is read. It may be that one newspaper has simply got the figures wrong but it is also possible that the papers who support the strike have picked the lowest from the mode, median and mean averages, whereas those who argue that the strike is unreasonable have picked the highest value. If the readers are told which type of average is being used there is no problem, but as long as the majority of the public are unaware of the existence of the mean, median and mode the term 'average' will continue to be used without qualification.

On its own, a measure of mean, median or mode often hides a lot of information which might be useful. For example, knowing that two batsmen have the same mean, median and mode scores might lead us to assume that they had very similar scoring records. With information only about 'average' scores, even if we know mean, median and mode, it is impossible to distinguish between the consistent batsman who has a mean average of 30 as a result of scoring 29, 30 or 31 runs in each match and the erratic scorer whose mean average of 30 is a result of several ducks combined with a few tremendous innings. On pages 14–19 you will learn how to summarise the variation in scores using concepts such as **standard deviation**. When you have learnt what standard deviation is you will be able to get a very full picture of a set of data from a brief summary. For example, if told that the mean for IQ test scores is 100 with a standard deviation of 15, you will immediately be able to work out that just over two-thirds of people have IQ scores of between 85 and 115 and that 95% of people's IQs range between 70 and 130. With the aid of a simple statistical table you will be able to say what percentage of the population score more or less than any IQ score you wish to quote.

You could, of course, do exactly the same without knowing about mean and standard deviations, but to do this it would take you very much longer since you would have to have a table containing all the individual results of IQ tests that have ever been taken. You would then have to read through them and, with the aid of your pocket calculator, work out all the above information. It would take you a matter of a few weeks or months!

Together with visual methods such as graphs, measures of 'average' and of score variation reduce masses of jumbled, confusing numbers into clear descriptions which are easy to follow and allow people with even very limited numerical ability to understand what the information means.

Inferential statistics

Sometimes we want to go beyond describing the world as it is – we want to make predictions about what will happen in the future. We usually base our predictions on what has happened in the past. We can predict with near certainty that the sun will come up tomorrow morning because it always has done so in the past. We might predict (though not with such confidence) that a certain horse will win this year's Grand National, on the basis of descriptive statistics that tell us how well it has done so far this season. Once we start making predictions, we are in the field of inferential statistics. It would be impossible to run a complex society like ours without some method of predicting things like how many young children, elderly people, criminals, cars, etc. there are likely to be in the next 5, 10, or 20 years, so that we can plan and build schools, hospitals, prisons and roads to cope with the demand when it arrives. We could predict using tea leaves or the stars, but most people prefer to use a method which gathers together the relevant information over the past few years and makes predictions based on those figures. That is not to say that this system is foolproof. Most people have a few horror stories up their sleeves about the local motorway which can't cope with the amount of traffic that wants to use it, or some other such blunder. It was once predicted that it would be impossible to walk the streets of London by the year 2000, not because of muggers, but because the number of horses in the city in the late nineteenth century was expanding so rapidly that the 'statistician' could see no way in which the manure from the many thousands of horses he predicted would be around at that time could be removed! If the assumptions on which your statistics are based prove to be unfounded or the situation changes then the predictions made by inferential statistics will prove false.

The main use of inferential statistics that we shall consider in this book relates to the results of our experimental work. Imagine that you had performed an experiment to investigate the hypothesis that a particular drug improved people's short-term memory; after testing 10 or 20 people you found that they did indeed remember more when they had taken the drug. On presenting the information to a colleague he says, 'Oh, you've just been lucky, the drug doesn't really have an effect, it's just chance, it's not a reliable result.' Although you don't agree with him, you can see that he *might* be right, so what do you do? One thing you could do is to perform the experiment again, but how many times will you have to repeat it before your sceptical colleague is convinced? Fortunately the statisticians can help us here. We do not need to repeat experiments over and over again in order to be reasonably sure that a result is reliable – we can use a statistical test to tell us how likely it is that our result was just due to chance so that we can make a reasonable guess about the outcome of the experiment if it were to be *exactly* replicated in the near future.

On pages 34–68 you will find instructions about how to perform statistical tests. No one will ask you to remember the formulae – you may even get a friendly mathematician or computer to work them out for you – but you should know when to use a particular test, and when the test has been performed you should be able to look up the test result in a table which will tell you whether the result of your experiment is one that you can rely on or not. Pages 24–32 will tell you how to do this.

Descriptive statistics

These statistics simply describe what you have found when performing a practical. They make no attempt to go beyond the data obtained, they make no predictions as to whether the results are likely to be similar if the practical is repeated, and they do not explain what caused the result. This type of statistic allows the results of an experiment to be shown in a way that can be easily understood.

Types of Descriptive Statistic

Raw score table   Frequency distributions   Ranked scores

Measures of dispersion   Graphs   Bar charts

Pie charts   'Averages'   Measures of correlation

Table of raw scores

The scores produced by subjects are simply put down as a list. If there are a lot of scores it may be difficult for the reader to extract information such as who scored best, worst, etc.

Ranked scores

These show the relationship between scores. Using ranked scores it

Table 1

SUBJECT	RAW SCORE	RANKED SCORE	
A	6	3	
B	3	1	
C	8	4.5	(tied raw scores
D	8	4.5	share the same
E	4	2	rank)
F	11	6	

can easily be seen which subjects scored highest, lowest or any other position.

You can see from Table 1 that B had the lowest and F had the highest score (see Appendix 1 for a full description of ranking procedures).

Graphs and bar charts

The important thing is that these are clear. Always label the axes and give an explanatory title.

Say we want to draw a graph showing the relationship between the time of day and the temperature. The temperature would go on the vertical axis. The temperature is called a **dependent variable** because its value depends on what time of day it is. If it is possible to identify a dependent variable when drawing a graph, it is customary to put it on the vertical axis.

Don't use line graphs in situations where other visual forms are more appropriate. A common mistake that students make is to present the scores of individual subjects as a line graph, but as you can see in Figure 3 the shape of the graph simply depends on which order you choose to put the individuals in. If you want to present this data visually, try using a bar chart, which shows the individual score without linking it to its neighbour.

Pie charts

These consist of a circle (or pie) which is sectioned to show the relative contribution of the parts which make up the whole. The pie

Figure 3

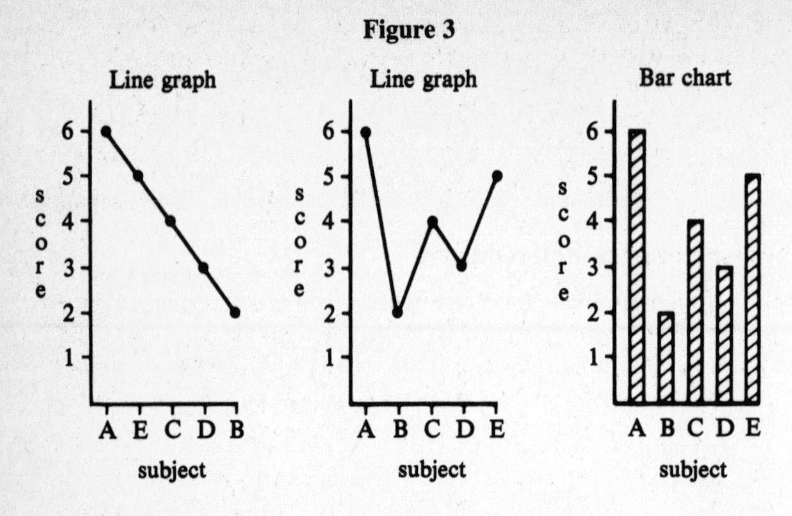

chart in Figure 4 shows the initial qualifications of an imaginary group of students in an A-Level Psychology class at a fictitious college.

This is very useful to give a generalised immediate impression, but not so useful to the reader who wants to know the exact

Figure 4

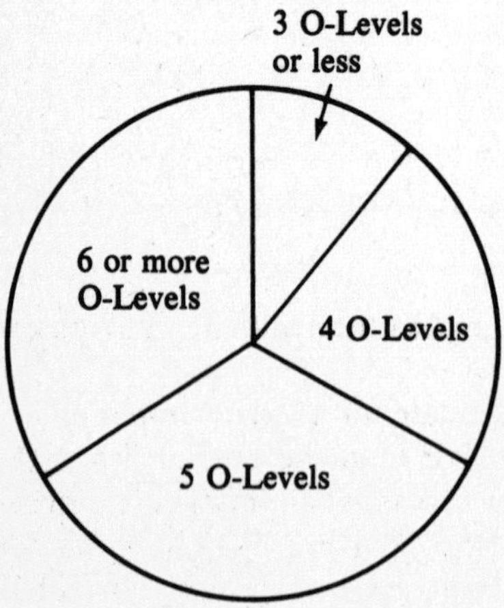

3 O-Levels
or less

6 or more
O-Levels

4 O-Levels

5 O-Levels

proportion of people in each category (though this information could easily be shown by inserting a percentage figure in each of the sections of the 'pie').

In Appendix 2 you will find a 'master' pie chart that has been split up into parts. By tracing the appropriate parts of this 'master' you can easily present your data in pie-chart form.

Measures of central tendency

It is often desirable to have one figure which can describe a group of scores. The layman uses the term **average** but this includes a multitude of different measures including **mean, median** and **mode.**

The *Mean* is the arithmetic average. It is found by adding up all the scores and dividing the result by the number of scores. For example,

5, 7, 4, 8, 6 MEAN = 30 ÷ 5 = 6

The *Median* is the middle score of a group of scores. This value has as many scores above as below it. It is found by placing the scores in order of size and finding the middle number. For example,

2, 3, 5, 6, 7, 9, 11, 13, 17 MEDIAN = 7

2, 2, 3, 4, 7, 8 MEDIAN lies between 3 and 4 = 3.5

The *Mode* is the score which occurs most frequently. For example,

2, 3, 3, 4, 6, 7, 2, 3, 3 MODE = 3

4, 7, 6, 3, 4, 9, 4, 4, 1 MODE = 4

Advantages (+) and disadvantages (−)

MEAN

(+) Takes into account the total and the individual values of all scores.
(−) Laborious to calculate for a large number of scores.
(−) Less representative than the median when the group of scores contains a few cases that are markedly different from the rest. For example, for the scores 2, 3, 4, 5, 6, 7, 57, the mean is 12, the median 5. Without the 'different' score of 57 the mean would be only 4.5.

MEDIAN

(+) More representative than the mean when there are extreme scores.
(+) Usually easier to calculate than the mean.
(−) It extracts less information than the mean since it does not use the precise numerical values of the scores.

MODE

(+) Easy to find.
(−) Very crude, especially if there is not much difference in the frequencies of the scores.

Frequency distribution curves

It is often useful to plot a graph which shows how frequently particular scores occur in your results.

For example, if you wanted to show how intelligence is distributed in the population at large, you would choose a lot of people at random and, assuming that IQ is a reasonable measure of intelligence, you would ask them to do an IQ test. You would probably present your results in the form of a frequency distribution graph. Since the subjects were chosen at random and there are many chance factors which may increase or decrease an individual's IQ, the graph you obtain would probably look like the one in Figure 5.

Figure 5

MEAN, MEDIAN AND MODE ALL HAVE THE SAME VALUE (AT POINT X)

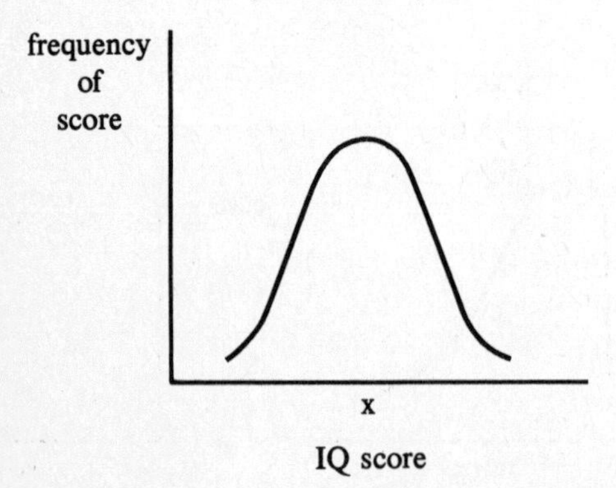

A graph which looks like this is called a **normal distribution**. It has the following three properties:

(a) Mean, median and mode occur at the same value.

(b) It is *bell-shaped* and has the same shape either side of the mean.

(c) The curve falls away relatively slowly at first on either side of the mean (i.e. many scores occur a little above or below the mean). Fewer and fewer scores are found with increasing distance from the mean.

A normal distribution is found when the following criteria are met:

(a) The data are continuous (like measures of height and weight which can vary by small amounts all the way along a scale, rather than separate or discrete measures like male/female or pass/fail in an exam where a person falls into one category or another).

(b) Each score is the result of a number of randomly distributed effects, some of which tend to increase the score whilst others decrease it. For example, your height is the result of a combination of many genes passed on from your parents, some of which tend to increase it and some to decrease it, together with environmental variables, such as the amount and type of food you eat, which may also increase or decrease height.

(c) There are a large number of scores drawn randomly from the population.

Normal distributions are not always found when frequency graphs are plotted. Sometimes distributions are *skewed*. The

Figure 6

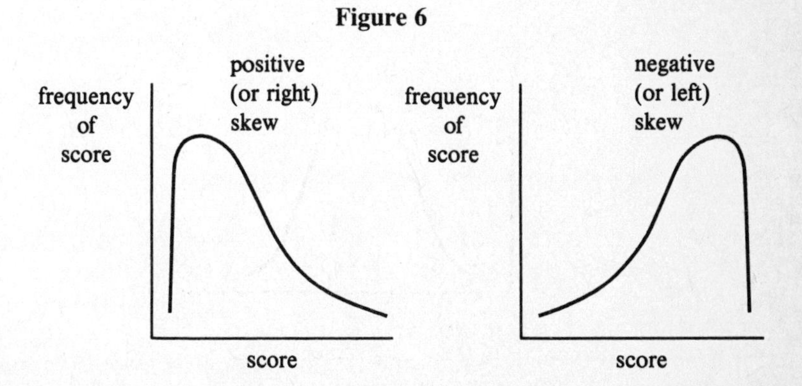

skewed distribution takes two forms – the median may have a higher or lower value than the mode (see Figure 6).

The characteristic of skewed distributions is that the mean, median and mode have different values.

A skewed distribution is often found when:

(a) A small number of scores are taken.

(b) A biased sample is taken of a population that may be normally distributed. For example, plot the frequencies of heights of males in the room in which you are sitting. Or, plot the distribution of university students' IQs.

(c) One end of the measuring scale has an attainable cut-off point. For instance, with **reaction time** (the time taken to react to a stimulus) there is no definite cut-off point for slow times since people can take as long as they like, but at the fast end of the scale it is not possible to better 0 seconds (a reaction time which is quite attainable, usually when the subject has anticipated the stimulus).

Measures of dispersion

These two sets of scores have the same mean:

21, 30, 45, 50, 60, 70, 85, 95 MEAN = 57
54, 55, 56, 57, 57, 58, 59, 60 MEAN = 57

Although these two groups of scores have the same mean, you can see that the scores in the first group are more widely spread out (dispersed) than those in the second. The mean cannot show the difference between these groups of scores. Some measure is needed to describe the variation in the data. There are a number of measures which do this but we shall concentrate on two, the **range** and the **standard deviation**.

Range is the difference between the lowest score and the highest.

21, 30, 45, 50, 60, 70, 85, 95 MEAN = 57 RANGE = 74
54, 55, 56, 57, 57, 58, 59, 60 MEAN = 57 RANGE = 6

A problem with the range is that, since it only takes into account the two extreme scores, it cannot give a good description of a group which has an odd score which is markedly higher or lower than the rest. For example,

2, 4, 4, 5, 7, 9, 10, 73 RANGE = 71

(This is not very representative of the group as a whole.)

Standard deviation is a more useful measure of distribution. It is a measure of the distribution of scores around the mean.

If the standard deviation of a group of scores is large, this means that the scores are widely distributed with many scores occurring a long way from the mean. If the standard deviation is small, most scores occur very close to the mean. (See Figures 7 and 8.)

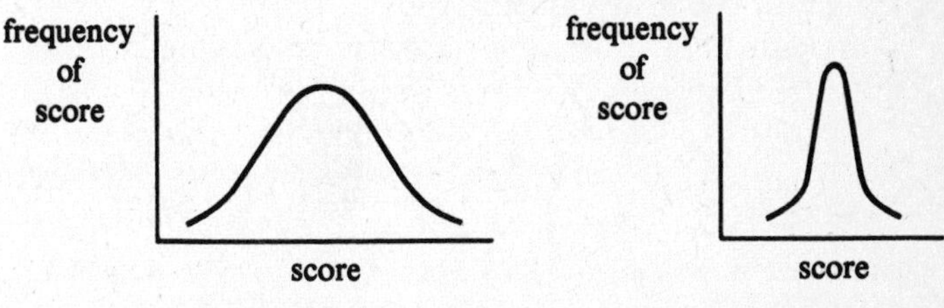

Figure 7
LARGE STANDARD DEVIATION

Figure 8
SMALL STANDARD DEVIATION

The lowest standard deviation is 0, indicating that there is no deviation at all (all the scores are identical).

If the scores are **normally distributed**, a knowledge of SD tells us what proportion of the scores falls within certain limits.

In fact, 68.26% of all scores lie between +1 and −1 SD from the mean – that is, between one standard deviation above the mean and one standard deviation below the mean. 95.44% of all scores lie between +2 and −2 SD from the mean. (See Figure 9 – we have rounded the numbers off to the nearest decimal point.)

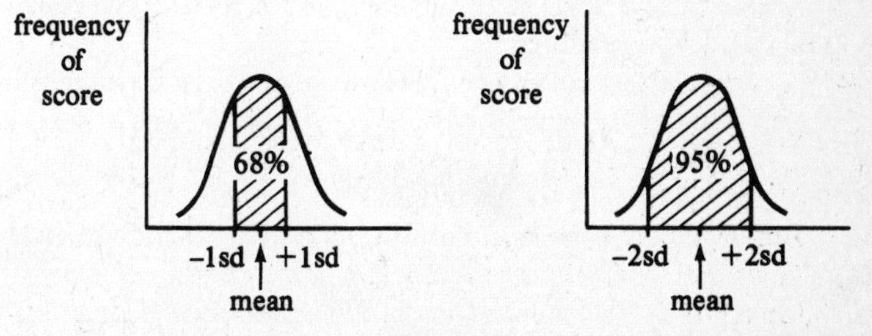

Figure 9

The value of SD is given in terms of the scores that you are plotting. For example, for IQ in the general population, the mean is 100 and SD is 15 (for most IQ tests). Therefore, 68% of the population scores between 85 and 115 IQ points (see Figure 10), and 95% of the population scores between 70 and 130 IQ points. Thus, from a knowledge that IQ is normally distributed with a mean of 100 and SD of 15, we can say that very few people (5% of the population) score above 130 or below 70.

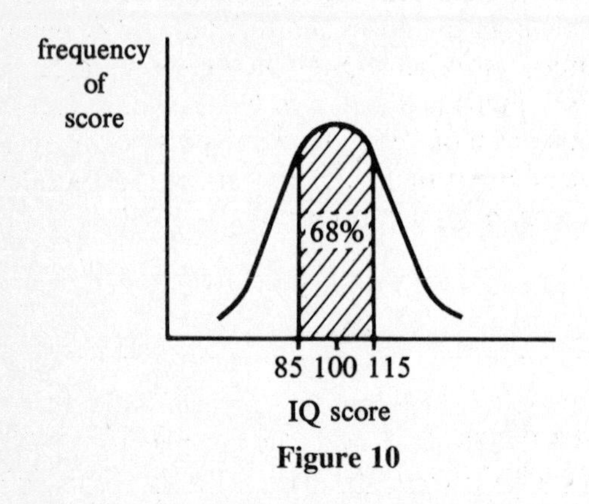

Figure 10

If you give IQ tests to a group of people and find that most of them show an IQ below 70, it is unlikely (drawing on your knowledge of mean and SD given above) that this has happened by chance sampling from the general population. These people have probably come together *because* they have low IQ. You can thus make inferences from the knowledge of the mean and SD of normally distributed scores.

QUESTION: If the mean score on a personality test is 12 and the SD is 4, between what two personality scores do 95% of people score (assuming normal distribution)?

ANSWER: Since 95% of the population score between plus and minus two standard deviations from the mean and in this case each standard deviation equals 4 personality points, then 95% of people score between 4 and 20 on this test.

See page 67 for the formula for working out standard deviation.

Standard deviation and Z scores

A knowledge of the mean and standard deviation of a group of scores allows us to convert any individual score into what is known as a **standard score** or **Z score**. The advantage of standard scores is that they allow us to pinpoint exactly how the individual compares with other people. Once again we will use IQ scores to demonstrate the use of standard scores, but the principle is the same for any test score as long as you know both the mean and the standard deviation of the scores. Remember that when IQ tests are given to large groups of randomly picked individuals, the resulting scores are normally distributed with a mean of 100 and a standard deviation of 15.

Z scores are simply a statement of how many standard deviations a subject's raw score is from the mean. A raw IQ score of 100, when

Table 2

Z SCORES

Z score	*Approx. % of scores above a positive or below a negative Z score*	*Z score*	*Approx. % of scores above a positive or below a negative Z score*
		1.5	7
0	50	1.6	5
0.1	46	1.7	4
0.2	42	1.8	3.6
0.3	38	1.9	2.8
0.4	34	2.0	2.3
0.5	31	2.1	1.8
0.6	27	2.2	1.4
0.7	24	2.3	1.0
0.8	21	2.4	0.8
0.9	18	2.5	0.6
1.0	16	2.6	0.5
1.1	14	2.7	0.4
1.2	12	2.8	0.3
1.3	10	2.9	0.2
1.4	8	3.0	0.1

converted to a Z score, becomes 0. Thus 115 IQ points is +1 Z score because 115 is one standard deviation above the mean.

Suppose we want to know what proportion of people scores less than Fred, who has an IQ of 85. This is what we do.

(a) First find out how many standard deviations Fred is from the mean. He is 15 points below the mean, therefore he is one standard deviation below the mean. This is known as a −1 Z score (the minus indicates that it is below the mean).
(b) Look up in the table above what percentage of the scores fall below a −1 Z score.
(c) You will find that 16% of the population score below a −1 Z score, so 16% of people have lower IQs than Fred.

If we want to find out the same information about Joe whose IQ is 80, we go through the same steps.

How many standard deviations is Joe from the mean? He is 20 points below the mean and every 15 points is one standard deviation. Therefore he is 1.3 standard deviations below the mean, so his Z score is −1.3. According to the table, only 10% of the population score less than this.

Often subjects are not so obliging as to gain scores which are so easily converted to Z scores as Fred and Joe. If this is the case, work out Z using the formula below:

$$Z = \frac{X - \bar{X}}{SD}$$

(X is the subject's score)
($\bar{X}$ is the mean)
(SD is the standard deviation)

Test yourself on the following example (answers at the bottom of the page).

A population of pygmies has just been discovered in the jungles of North Cheshire: they have a mean height of 3 feet and a standard deviation of 6 inches.

(a) What percentage of the pygmy population is shorter than 2 feet 6 inches?
(b) What percentage is shorter than 2 feet 3 inches?
(c) What percentage is shorter than 2 feet?

Answers: (a) 16% (b) 7% (c) 2.3%.

Since a normal distribution curve is symmetrical, you will have already realised that the table can be used to work out the percentage of scores above a score that is above the mean. For example, an IQ of 115 (+1 Z score) has 16% of scores above it.

Z scores allow us to make comparisons between performances on different tests. If, for example, a subject gained 25 marks on a Maths test and 45 on an English test, this does not necessarily mean that he is better at English than Maths; even if both tests were marked out of the same total it might simply be that the Maths test was harder or that the English teacher was a more lenient marker. The calculation of standard scores (Z scores) for the subject on each test eliminates the effect of these differences. If the subject's Z score is greater on the English test than the Maths test, then he really has done better at English because we are comparing him with other people who took the tests – the Z scores tell us that he is better at English compared with other people than he is at Maths.

Correlation

We often want to find the relationship between two sets of variables. For example, we might want to know whether there was a relationship between the amount of television that people watch and their intelligence. To do this we could ask people to record the amount of TV they watched over a particular period and also give them an IQ test. The results could be shown as in Figure 11 in the form of a scattergram, where each point represents one person's score on the two variables (IQ and amount of TV watched).

Scattergrams to show possible results of a correlational study of the relationship between IQ and TV watching.

Figure 11 shows a strong relationship between the two variables. Those subjects with the highest IQ watch the most TV. This type of relationship, where people who score highly on one variable tend also to do well on the other, is known as a **positive correlation**.

Figure 12 is another possible result which also shows a strong

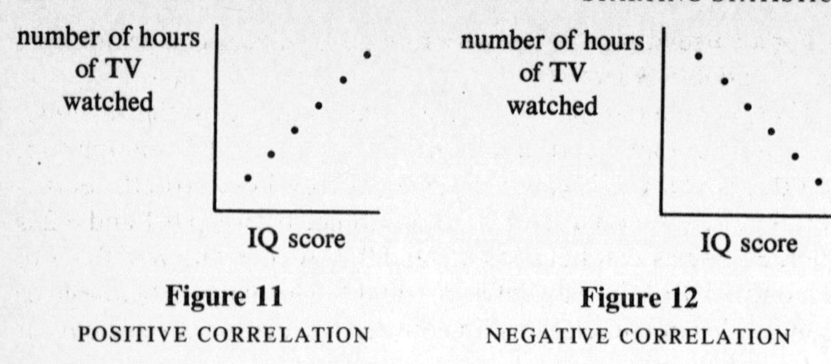

Figure 11
POSITIVE CORRELATION

Figure 12
NEGATIVE CORRELATION

relationship between the two variables, but this time people with higher IQs tend to watch least TV. This type of relationship, where people who score highly on one variable tend to have a low score on the other, is known as a **negative correlation**.

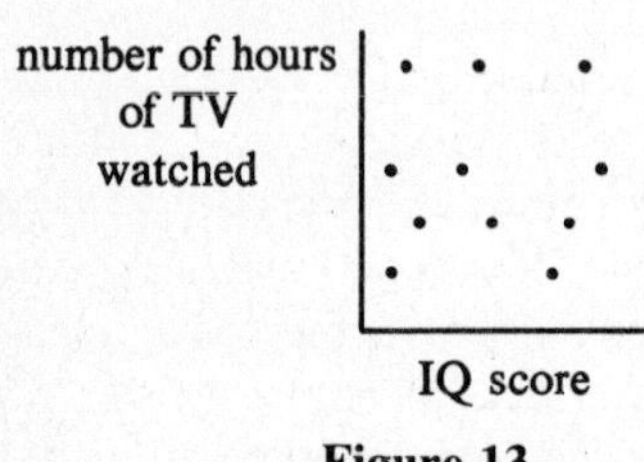

Figure 13
ZERO CORRELATION

In practice it is unlikely that we should find a very strong relationship between these two variables. The relationship shown in the scattergram in Figure 13 is near zero.

Producing a scattergram from your data will give a quick visual impression of the relationship, but a test of correlation is required to provide a precise numerical value. Spearman's rho is an example of such a test which tells you the extent to which two variables correlate – it gives you the correlation co-efficient (which may be anything between +1 and −1). The closer the co-efficient to +1 or −1, the more perfect the relationship; the closer it is to 0, the weaker is the relationship.

For a correlation of +1, every rise in variable A is reflected by a rise in variable B.

For a correlation of 0, there is no relationship between A and B.

For a correlation of −1, every rise in variable A is reflected by a fall in variable B.

Many people misinterpret correlation co-efficients: they assume, for example, that a correlation of +0.5 means a 50% relationship. But this is not true because the scale of correlation co-efficients is not like the scale on a ruler – the difference between 0.1 and 0.2 is not the same as that between 0.8 and 0.9. A correlation of 0.5 is in fact only 25% of the way between 0 and 1. The diagram below gives a rough idea of the relationship between correlation co-efficients of different values.

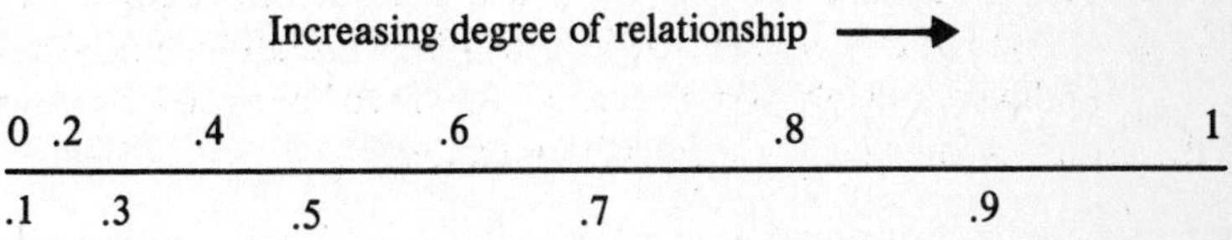

Increasing degree of relationship →

0 .2 .4 .6 .8 1

.1 .3 .5 .7 .9

As you can see, correlations of 0.1 and 0.2 are closer to no relationship and 0.9 is further from a perfect relationship than might have been thought. To get an approximate idea of where a co-efficient lies on the scale between 0 and 1, square the number immediately after the decimal point. Thus a correlation of 0.3 is 9% of the way along the scale, and 0.9 is 81% of the way.

Once you have found the correlation co-efficient for your data, you must test it for **significance** – this tells you the probability of the co-efficient being due to chance rather than to a consistent relationship between the variables. If a correlation co-efficient is significant, it is unlikely to have occurred by chance and will probably be found again if you take measures of the same variables from a similar sample of people. Table 21 on page 62 is used to work out whether a correlation is significant; have a look at the table and note how even a small correlation may be significant if it has been gained from a large sample, but that even a correlation of +1 is probably due to chance with only a couple of pairs of scores. This is because chance is more likely to influence results when only a small number of observations have been made.

A very common mistake is to assume that because two variables correlate, one causes the other. It is vital that you do not make this assumption, because:

correlation does not imply causation.

There is a correlation of +0.95 between the length of railway line in a country and the incidence of certain types of cancer. This does not mean that railways cause cancer – there may be some underlying cause which affects both factors, such as industrialisation. If two variables correlate, one *may* cause the other but the correlational technique cannot confirm this.

Only the experimental method can study cause and effect. This is because it manipulates independent variables whilst controlling other variables in order to discover the effect on dependent variables (see page 27).

Tests of correlation such as Spearman's rho only detect simple linear relationships. If the relationship is like the one in the scattergram shown in Figure 14, a correlation of near zero will be found because the first half shows a positive relationship and the second half a negative. Always draw a scattergram before calculating the correlation co-efficient just in case your variables have one of these more complex relationships.

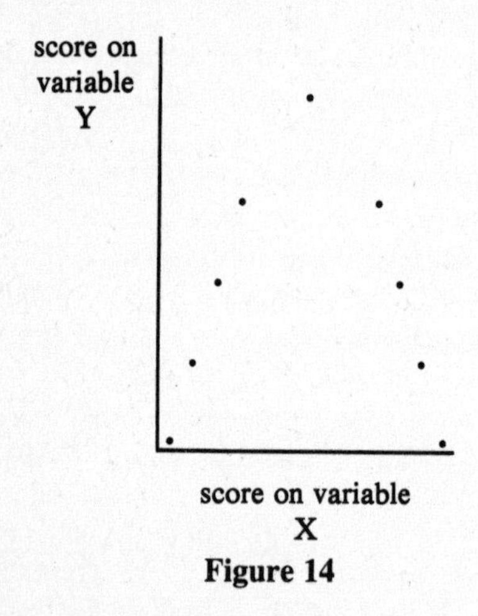

Figure 14

Uses of correlation

(1) Correlation can be used to test the reliability or validity of a measuring instrument such as a questionnaire or a selection procedure. For example, if the scores on a test one day have a high (near +1) positive correlation with those on another day,

the test is reliable. (See the section on reliability and validity, page 78.)

(2) If two variables are highly correlated, we can make predictions from one to the other. For example, if a selection test score correlates highly with a measure of performance on a job, it can be used to choose between candidates since we assume that those who do well on the test will perform the job well.

(3) In some areas of study, correlational studies represent the only way of getting worthwhile information since an experiment may not be possible without producing a very artificial situation or may be undesirable for ethical reasons. This is particularly true in clinical, educational and industrial psychology.

(4) Correlational studies frequently suggest hypotheses that may later be tested by means of an experiment.

Experiments and inferential statistics

Psychologists perform inferential statistical tests to find out whether the results of their experiments occur because of chance factors – that is, if the experiment were repeated would a similar answer be found? For example, if we performed an experiment and found that students who had gone through a course using programmed learning books did better than those using ordinary books, statistical tests would help us to decide whether this difference was due to the different type of book or simply to chance factors which might lead to a different result if the experiment were repeated. (We make an *inference* from our data as to whether the result is likely to be reliable or not.)

In different situations, different statistical tests are used. In order to know when to use which test, we must first know something about:

(a) types of data
(b) the definition of an experiment
(c) experimental design.

Types of data

The data that we collect from experiments and observations may take different forms: sometimes we just count, other times we record the order of things and at other times we measure using a variety of different implements (rulers, clocks, etc.).

Nominal data (counting)

Whenever you simply count the number of subjects that did one thing or another, or fall into this category or that, you are using **nominal** (sometimes called **frequency**) **data**. For example, if I say that there are four women and five men in a room, this is nominal data since it tells you the frequency of occurrence of instances of the categories men and women.

Information that tells us such things as how many people saw a film last night or that it rained on twenty and was dry on ten days last month is all nominal data.

Nominal data also includes numbers used simply as labels, such as those found on buses and sports competitors' shirts.

Ordinal data (ordering)

The results of many sporting events are given in the form of **ordinal** (sometimes called **ranked**) **data**. You are told who came first, second, etc. For example, if I say that in the National, Dobbin came first (1), Neddy came second (2) and Blue Donkey came third (3), this is ordinal data. It does not tell you how much difference there was between the horses, simply the order in which they finished.

Interval and ratio data (measuring)

Perhaps the most common form of data is derived from the use of measuring instruments such as clocks, weighing scales and thermometers. These instruments give data in the form of 'public units of information' such as seconds, minutes, pounds, ounces, kilograms, degrees Celsius and Fahrenheit. They are called public units of information because each unit has an agreed value; because each unit has an agreed value the difference between 4 and 5 pounds is exactly the same as that between 10 and 11 pounds, just as the difference between 6 and 7 seconds is the same as that between 8 and 9 seconds.

Interval and **ratio data** give more than just order; they also show how much difference there is between the first and the second, the second and the third, etc. If I have three sticks which are 6, 4 and 2 inches long, you know not only which sticks are the longest but also by how much one stick is longer than another, because I have described the sticks using ratio data. If your data describes each value in terms of the exact number of minutes, seconds, feet, inches, pounds, ounces, degrees or any other units of public information, it is interval or ratio data.

These two types of data differ in only one respect: ratio data has a logical zero point whereas zero on an interval scale is a purely arbitrary point. If you get a reading of 0 on a weighing machine, ruler or clock it means that there is no weight, length or time measured; as long as the machines are working properly there can be no argument about this, as you are dealing with a ratio scale of measurement. Interval scales are less common but the classic example is temperature where a figure of 0 does not mean there is no temperature (after all 0°C is the same as 32°F). It may not seem worth making this distinction between interval and ratio data, but it is important for mathematics since it is quite reasonable on a ratio scale to say that 4 (pounds of potatoes, perhaps) is twice as much as 2 but it is not reasonable to make the same argument on an interval scale. For example, 4°C is *not* twice as hot as 2°C (after all, if you had been using a Fahrenheit thermometer the temperatures would have been 35.6°F and 39.2°F).

The statistical tests that you are likely to come across do not distinguish between interval and ratio data, so in practice there is no need to worry about whether your data is one or the other.

Here is an example of the use of the three types of data used to describe four lines – A, B, C and D.

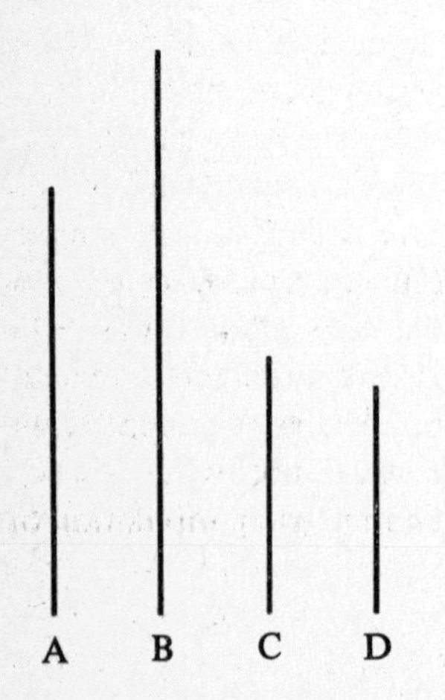

NOMINAL DATA DESCRIPTION

Two lines are longer than one inch. Two are shorter than one inch.

ORDINAL DATA DESCRIPTION

Line	Rank
A	3
B	4
C	2
D	1

RATIO DATA DESCRIPTION

Line	Length (ins.)
A	1.5
B	2.0
C	0.9
D	0.8

As you can see from the description of the lines, *ratio data contains most information, nominal contains least.*

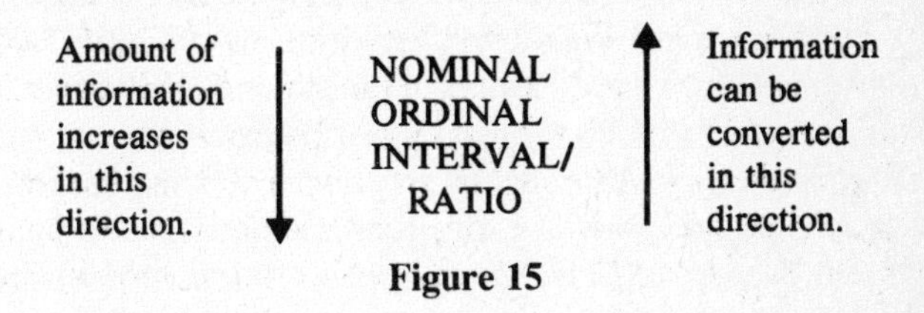

Figure 15

By discarding some information it is possible to change ratio data into ordinal or nominal data. It is not possible to change ordinal or nominal into ratio without getting more information. (See Figure 15.)

Definition of an experiment

An experiment is a study of *cause* and *effect*. It differs from simple observation in that it involves deliberate manipulation of one variable (the **independent variable**), while controlling other variables (**extraneous variables**) so that they do not affect the outcome, in order to discover the effect on another variable (the **dependent variable**). Simple observation merely involves observing what is going on without attempting to produce change by manipulation of variables.

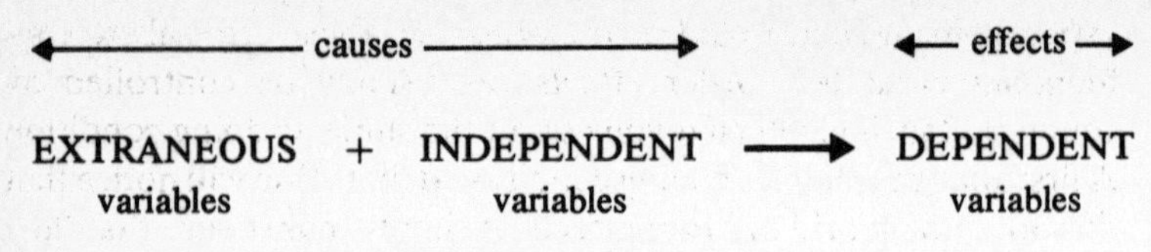

Figure 16
FORMULA FOR AN EXPERIMENT

All experiments can be fitted into the 'formula' shown in Figure 16. The independent variable is the one that the experimenter has decided to vary in the different conditions. The results that the experimenter collects are measures of the dependent variable. In an experiment to test the hypothesis that there is a difference in the speeds with which people react to visual and auditory stimuli, the independent variable would be whether the stimulus was a sound or a light, and the dependent variable would be the time taken to respond to these stimuli. In an experiment to test the effect of a drug on memory, the amount of drug received by the subject would be the independent variable and the dependent variable would be some measure of memory. It is not enough, however, to simply think about the independent and dependent variables when designing an experiment. There are lots of other things which might affect the results, and unless these are controlled in some way to ensure that any effect they have is the same in both conditions, we have no way of knowing whether any difference in the dependent variable was due to the independent variable or to one of these other (extraneous) variables.

Examples of extraneous variables and their control. In the auditory and visual reaction time experiment above, you would probably decide to use the same people in both conditions; if you used different people in each condition any difference in the results might simply be due to individual differences between subjects. Using the same people to react to both visual and auditory stimuli, however, introduces another set of extraneous variables, namely order effects. If all subjects responded to the auditory stimulus first, then it could be argued that any difference in the results of the two conditions was due to practice (if visual reaction time was best) or fatigue (if auditory reaction time was best). This would be a confounding nuisance when it came to interpreting the results of the experiment; uncontrolled extraneous variables are called **confounding variables**. Order effects can usually be controlled by counterbalancing, which involves half the subjects doing condition A first and the other half doing condition B first. You will notice that this does not get rid of order effects, it simply makes sure that they have the same effect on each condition.

In all experiments the subjects should, as far as possible, be treated the same way in each condition, and care should be taken to avoid such things as competition between subjects or letting the subject know how you expect the experiment to turn out – unless, of course, you are investigating the effects of this.

It is impossible to control all extraneous variables, but we must design our experiments to avoid any constant difference between the conditions other than the independent variable. Some variables change in a random manner – these are chance factors and cannot be controlled because we can't predict them. Chance variables make life interesting and are the basis of gambling but they are a pain in the neck for experimenters. It is these uncontrollable factors that make statistical tests necessary; tests are used to estimate how likely it was that any difference between the results of the experimental conditions was due to chance. (See page 91 for a list of methods used for control of extraneous variables.)

EXTRANEOUS VARIABLES	+	INDEPENDENT VARIABLES	→	DEPENDENT VARIABLES
		Condition A		Result A*
		Condition B		Result B*

*The difference here is due to the difference in the variables on the left. If *extraneous* variables were controlled to prevent *confounding* variables, then this difference is due to the difference in the *independent variable* or to *chance*.

Experimental design

For the purpose of the tests that we shall use, we must know about three types of experimental design. Each design uses subjects in a slightly different way.

(a) **Repeated measures design**. In this design, each subject performs in both the control and the experimental condition (i.e. in both condition A and condition B).

(b) **Independent subjects design**. In this design, *some* subjects perform in condition A and *others* in condition B. (Subjects must be allocated randomly to each group.)

(c) **Matched pairs design**. From the results of a pretest, the subjects are sorted into matched pairs (pairs of equal abilities on the task to be measured). One from each pair performs in the experimental condition and one in the control.

Table 3

AN EXAMPLE OF ALLOCATION OF SUBJECTS TO CONDITIONS IN REPEATED MEASURES AND INDEPENDENT SUBJECTS DESIGNS

(a) REPEATED MEASURES		*(b) INDEPENDENT SUBJECTS*	
Condition A	*Condition B*	*Condition A*	*Condition B*
Matilda	Matilda	Matilda	Joe
Jeremy	Jeremy	Jeremy	Freda
Joseph	Joseph	Joseph	Frank
Mary	Mary	Mary	Anita
Joy	Joy	Joy	Christine

Advantages and disadvantages of the designs

(1) As you can see, the independent subjects design is quite wasteful with subjects compared with repeated measures.

(2) Since the control group is the same as the experimental group in repeated measures design, this method automatically ensures that there are no personality or ability differences between the two groups.

(3) In some experiments, performance in one condition 'pollutes' the subject for use in the other condition, since he is no longer a naive subject. This is often the case in learning experiments. For example, it would not be possible to use repeated measures design to investigate different methods of learning to drive a car.
(4) Matched pairs design controls for personality and ability differences between conditions by the method of constancy (see page 91), and can be used in situations where repeated measures design is not possible.
(5) The choice of characteristics to match in matched pairs design is a subjective decision and pretesting can take a long time. Independent subjects design is more commonly used in situations where repeated measures is not appropriate.

Why do we use inferential statistical tests?

Statistical tests tell us the likelihood of obtaining the results we get simply as a result of chance factors (small extraneous variables, too unpredictable to control). If the test says that it is unlikely that the difference we found in the *dependent variable* was due to chance, this means the difference was probably due to the difference in the *independent variable*, and that if the experiment were repeated we would probably get the same sort of result (i.e. if condition A produced a 'better' result than condition B, this direction of difference would be repeated).

If the test tells us the result was *probably not due to chance*, the difference is said to be *significant*.

If the test tells us the results were *probably due to chance*, the difference is said to be *not significant*.

A '*significant*' difference is probably repeatable. A '*not significant*' difference may occur in the opposite direction or not occur at all if the experiment is repeated.

Level of significance

The **level of significance** is normally taken at the 5% level of probability – this means that there is only a 5% probability that the result occurred due to chance. The difference is then said to be significant if there is only a 5% (or less) probability that it occurred due to chance.

(Please refer to page 69 for further discussion of the concept of significance.)

Now that you have read this section you can use the chart set out in Table 4 to decide which test to use for your present practical.

Table 4

CHART FOR CHOOSING TESTS OF SIGNIFICANT DIFFERENCE

	TYPE OF DATA		
DESIGN	NOMINAL	ORDINAL	INTERVAL/RATIO
REPEATED MEASURES	Sign test	Wilcoxon	Related t
MATCHED PAIRS	Sign test	Wilcoxon	Related t
INDEPENDENT SUBJECTS	Chi squared	Mann-Whitney	Unrelated t

N.B. See page 75 for a discussion of the assumptions made by parametric tests such as the related and unrelated t tests.

Because interval/ratio data can be easily converted into ordinal or nominal data, tests which only require nominal data can cope with interval/ratio data. Tests which require ordinal data can also cope with interval/ratio data. Converting from interval/ratio to other kinds of data does involve throwing away some information and so you will be slightly less likely to find a significant difference.

Statistical tests – formulae and examples

Sign test

Experimental design: Repeated measures or matched pairs.
Minimum level of data: Nominal.
What the test does: The test simply counts the number of times one condition is larger than the other and compares this with what would be expected by chance if there was no real difference between the conditions.

(1) Count the number of subjects left after discarding those with the same score in condition A and condition B. Call this total N.

(2) Look at each person's score in both conditions. If they have a larger score in condition A than they have in condition B, put a plus (+) sign next to their scores. If they have a larger score in condition B than they have in A, put a minus (−) sign next to their scores.

(3) Count the least frequent sign. Call the total of the number of these signs X.

(4) Use Table 5: if X is *equal to or lower than* the value in the table which is next to your N, then the result is significant at the 5% level for a *two-tailed test* (see page 74).

Table 5

N	X
6	0
7	0
8	0
9	1
10	1
11	1
12	2
13	2
14	2
15	3
16	3
17	3
18	4
19	4
20	5
21	5
22	5
23	6
24	6
25	7

N=the number of pairs of scores after discarding those with equal scores in both conditions

Note. The test cannot be used when there are fewer than 6 pairs of scores.

Example

An example of the use of the sign test is on the results of a practical to investigate the experimental hypothesis that there will be a significant difference between subjects' scores on an arithmetic test when performed after drinking four pints of beer and scores obtained after no alcohol consumption.

Table 6

RESULTS

SUBJECT	SCORES FOR CONDITION A (no beer)	SCORES FOR CONDITION B (with beer)	STEP 2
1	70	67	+
2	62	63	–
3	57	52	+
4	77	72	+
5	81	81	
6	45	44	+
7	27	22	+
8	63	60	+
9	65	65	
10	44	39	+
11	39	38	+
12	36	34	+

STEP 1 The number of subjects left after ignoring subjects 5 and 9, who did equally well in the two conditions, is 10. Therefore N = 10.

STEP 2 For each subject a '+' is recorded if their largest score was in condition A and a '–' is recorded if their largest score was in condition B. See Table 6.

STEP 3 The least frequent sign is '–', which occurs once, so X = 1.

STEP 4 The table shows that when N = 10 an X value of 1 denotes a significant difference at the 5% level, so we can reject the null hypothesis and accept the experimental hypothesis. Provided that the experiment was well designed with no confounding variables, the results show that beer drinking does impair performance on this type of task. (N.B. We are, of course, using fabricated data in this example!)

Wilcoxon test

Experimental design: Repeated measures or matched pairs.
Minimum level of data: Ordinal.
What the test does: The test examines the differences between subjects' scores in each of the conditions. It looks at the direction of the difference and also ranks them to compare the relative size of those differences where subjects do best in condition A and those where performance in condition B is best. If chance was responsible for the difference between conditions A and B, we would expect the number and relative size of differences in favour of condition A to be similar to those in favour of condition B.

(1) Discard the results of any subject who scored the same in both conditions.

(2) Work out the difference between the two scores of each subject (the largest of the two scores minus the smallest).

(3) Rank the differences, giving the smallest a rank of 1 (see page 110 if you are unsure about ranking).

(4) Add up all the ranks for the subjects who did best in condition A; then add up all the ranks for those who did best in B.

(5) The *smaller* of these two figures is known as T.

(6) Look up the T score in Table 7. If your value is equal to or lower than the value in the 5% column, your result is significant at 5%. If it is *equal to or lower than* that in the 1% column, it is significant at 1% (even less likely to have occurred due to chance).

Example

Here is an example of the use of the Wilcoxon test. In order to show the differences between this and the sign test, we are working with the same data as we used on page 36. (See Table 8.)

Table 7

	SIGNIFICANCE LEVEL (TWO-TAILED)	
N	*5%*	*1%*
6	0	–
7	2	–
8	4	0
9	6	2
10	8	3
11	11	5
12	14	7
13	17	10
14	21	13
15	25	16
16	30	20
17	35	23
18	40	28
19	46	32
20	52	38

N=number of subjects (after discarding those who did equally well in both conditions)

Source: Adapted from F. Wilcoxon, *Some Rapid Approximate Statistical Procedures* (American Cyanomide Co., 1949), Table 1.

STEP 1 The number of subjects left after ignoring subjects 5 and 9 who did equally well in the two conditions is 10. Therefore N = 10.

STEP 2 See Table 8.

STEP 3 See Table 8.

STEP 4 The total ranks for the subjects who did best in condition A = 53. The total ranks for those who did best in condition B = 2.

STEP 5 T is the smaller of the above totals, therefore T = 2.

STEP 6 The critical value for T when N = 10 is 8 for a 5% level of significance and 3 for the 1% level. Our value for T is 2. Since this is smaller than either of the critical values

Table 8

RESULTS

SUBJECT	SCORES FOR CONDITION A (no beer)	SCORES FOR CONDITION B (with beer)	STEP 2 DIFFERENCE BETWEEN A AND B	STEP 3 RANKINGS
1	70	67	3	5.5
2	62	63	1	2*
3	57	52	5	8.5
4	77	72	5	8.5
5	81	81		
6	45	44	1	2
7	27	22	5	8.5
8	63	60	3	5.5
9	65	65		
10	44	39	5	8.5
11	39	38	1	2
12	36	34	2	4

(* indicates the rank for a subject who did best in condition B)

above, the difference between the scores in the two conditions is significant at the 1% level, so the null hypothesis may be rejected and the experimental hypothesis accepted.

The raw data used for this example is the same as that for the sign test example (page 36). When the sign test was used we found the result significant at only the 5% level; the Wilcoxon test however takes into account not just the number of times one condition is found to be greater than another (nominal data) but also the ranks of the *size* of the differences between the conditions (ordinal data). So although both tests can be used for repeated measures design, the Wilcoxon test, because it looks at both the direction and the size of the differences, is the more sensitive of the two; that is, it is more likely to reveal a significant difference. The related t test (see page 54) is even more sensitive to significant differences because it takes into account the actual size of the differences between the conditions (interval/ratio data) rather than just the ranks of these differences.

Mann-Whitney test

Experimental design: Independent subjects.
Minimum level of data: Ordinal.
What the test does: The test looks at the size of the scores in condition A and condition B. If most of the large scores are in one condition, this reduces the likelihood that the difference between the two conditions was due to chance alone.

(1) Rank scores as if they come from a single group, with the lowest score receiving rank 1 (see page 110 for help with ranking).

(2) Add up the ranks assigned to the smallest group (if they are both the same size use either group).

(3) Substitute into the following formula:

R = total of the ranks in the smallest group
N_1 = number of cases in the smallest group
N_2 = number of cases in the largest group

$$U_1 = N_1 N_2 + \frac{N_1(N_1+1)}{2} - R$$

$$U_2 = N_1 N_2 - U_1$$

(4) Take the *smaller* of U_1 and U_2 and if this U value is *equal to or lower than* the critical value in Table 9, then the difference is significant at the 5% level for a two-tailed test.

To find the critical value in Table 9, look down the column headed by the number of subjects in your smallest group and across the row next to the number in the largest group. The critical value is at the point were the row and column meet.

Table 9

N2 \ N1	6	7	8	9	10	11	12	13	14	15	16	17	18	19	20
7	6	8													
8	8	10	13												
9	10	12	15	17											
10	11	14	17	20	23										
11	13	16	19	23	26	30									
12	14	18	22	26	29	33	37								
13	16	20	24	28	33	37	41	45							
14	17	22	26	31	36	40	45	50	55						
15	19	24	29	34	39	44	49	54	59	64					
16	21	26	31	37	42	47	53	59	64	70	75				
17	22	28	34	39	45	51	57	63	67	75	81	87			
18	24	30	36	42	48	55	61	67	74	80	86	93	99		
19	25	32	38	45	52	58	65	72	78	85	92	99	106	113	
20	27	34	41	48	55	62	69	76	83	90	98	105	112	119	127

Source: Adapted from D. Auble, 'Extended tables for the Mann-Whitney statistic', *Bulletin of the Institute of Educational Research at Indiana University*, vol. 1, 1953, number 2.

N.B. If a large number of subjects have the same scores the test becomes less sensitive to significant differences and a type 2 error (see page 71) is more likely. The effect is small, but consider using a different test if there are more than 25% shared scores.

Example

Here is an example of the use of the Mann-Whitney test on the results of an experiment to test the hypothesis that there is a significant difference in the number of days taken to reach proficiency on a computer-user test using two different training packages.

The subjects in condition A used training package 1, those in condition B used training package 2.

Table 10

CONDITION A			CONDITION B		
	Time	*STEP 1*		*Time*	*STEP 1*
Subject	*(days)*	*Rank*	*Subject*	*(days)*	*Rank*
1	14	2	11	39	17
2	27	11	12	26	9.5
3	13	1	13	35	16
4	18	3	14	44	19
5	22	4	15	40	18
6	30	13	16	26	9.5
7	29	12	17	31	14
8	25	8	18	32	15
9	23	6	19	23	6
10	23	6	20	47	20
		—			—
	Total =	66 (STEP 2)			

STEP 1 The scores are ranked as if they come from a single group. See Table 10. If you have any difficulty with ranking refer to page 110.

STEP 2 There are equal numbers in each group; we will use condition A ranks to calculate R. The total of the ranks in condition A = 66 (see Table 10).

STEP 3 R = 66 N1 = 10 N2 = 10

$$U_1 = N_1.N_2 + \frac{N_1(N_1 + 1)}{2} - R$$

$$U_1 = 10\times10 + \frac{10(10 + 1)}{2} - 66$$

$$= 100 + \frac{110}{2} - 66$$

$$= 155 - 66$$

$$\underline{U_1 = 89}$$

$$U_2 = N_1.N_2 - U_1$$

$$= (10 \times 10) - 89$$

$$= 100 - 89$$

$$\underline{U_2 = 11}$$

STEP 4 U_2, being the smaller of U_1 and U_2, is the one to be compared with the critical value in the table.

The critical value when N1 = 10 and N2 = 10 is 23. Since U is lower than the critical value, the result is significant at the 5% level. We can, therefore, reject the null hypothesis (that there is no real difference between the training packages) and report that proficiency was achieved significantly sooner with training package 1.

Chi squared [χ^2] test

Experimental design: Independent subjects.
Minimum level of data: Nominal.
What the test does: The test looks at the number of observations made in each category and compares this with the number of observations which would be expected if there was no relationship between the variables and differences between the proportions in each category were simply a result of chance.

(1) Place your data into a 'box' table like the one shown in Figure 17 (the number of boxes depends upon how many variables you have studied). The example on page 47 gives hints on constructing this sort of table.

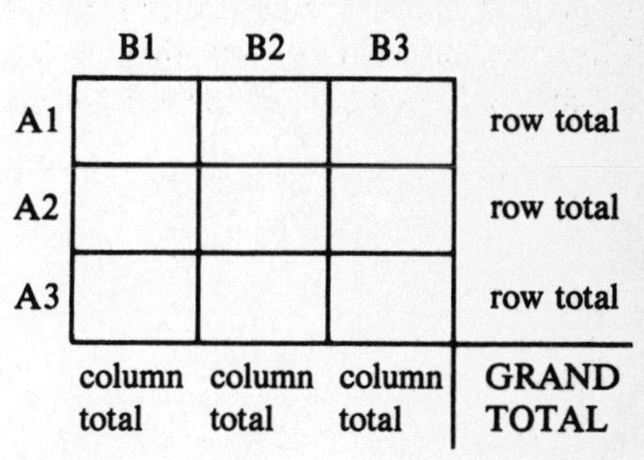

Figure 17

Each 'box' should contain an observed frequency (O), i.e. the number of observations that fit into that particular category. For example, the top left-hand box might record that 200 people from grammar schools were at present employed in a manual occupation and the box below would record the number of ex-comprehensive pupils in the same type of work. It is important that any one observation can be included in only one category.

The test compares the observed frequencies (O) with the frequencies expected if there was no relationship between the variables (E).

(2) Add up the figures in each row and column and place them as shown in Figure 17. The grand total is the total number of observations made.

(3) Work out the expected frequencies (E) for each 'box', using the following formula for each one.

$$E = \frac{\text{Row total} \times \text{column total}}{\text{Grand total}}$$

(4) Find the degrees of freedom (df) using the formula df = (number of rows − 1) × (number of columns − 1).

(5) If df = 2 or more, work out value of chi by using the following formula:

$$\chi^2 = \sum \frac{(O-E)^2}{E}$$

The easiest way to do this is to use a table like the one below and fill in each column in turn, adding up the final column to discover the chi squared value:

Box	Observed frequency	Expected frequency	O−E	$(O-E)^2$	$\frac{(O-E)^2}{E}$
Category 1					
Category 2					
Category 3					
etc.					

(6) If df = 1, an adjustment known as **Yates correction** should be applied and chi is calculated using the formula:

$$\chi^2 = \sum \frac{(|O-E|-\frac{1}{2})^2}{E}$$

N.B. $|O-E|$ means that you only work out the difference between O and E and ignore whether the result of O−E is positive or negative.

A similar table to the one above can also be used for this formula but an additional column is needed:

Box	O	E	O−E	$(\lvert O-E\rvert-\frac{1}{2})$	$(\lvert O-E\rvert-\frac{1}{2})^2$	$\frac{(\lvert O-E\rvert-\frac{1}{2})^2}{E}$

(7) Use the chi squared table on page 50 to see whether the result was significant. You will need the chi and df values. Note that the chi squared test cannot be used if any of the expected frequencies are less than 5. If you find that any expected values fall below 5, try combining some of the categories or go out and get some more data.

Example

Here is an example of the use of a chi squared test, to investigate the preferences for employment in certain industries by male and female school leavers.

RESULTS Out of 150 boys interviewed, 80 expressed a preference for the construction industry, 20 for textiles and 50 for banking. The preferences of the 150 girls were: 9 for construction, 80 for textiles and 61 for banking.

STEP 1 The results are placed into a box table (see Table 14). Note how the table is constructed by placing one set of variables across the top and the other set down the side so that each 'box' refers to a specific category, for example males choosing the construction industry (box A). Each 'box' has been labelled A, B, C, etc. so that it can be easily referred to.

Table 14

	Construction	Textiles	Banking
Male	A 80	B 20	C 50
Female	D 9	E 80	F 61

STEP 2 The numbers in the male row are added together to give a row total. The same is done for the females (see Table 15). The numbers in each of the industry columns are added together to give a set of column totals. The grand total is the total number of males and females, i.e. 300. Note that if you have done everything correctly, adding all the row totals together should give the grand total and so should adding all the column totals.

Table 15

	Construction	Textiles	Banking	
Male	A 80	B 20	C 50	150
Female	D 9	E 80	F 61	150
	89	100	111	300

STEP 3 The expected frequency for each 'box' is calculated using the formula:

$$E = \frac{\text{Row total} \times \text{column total}}{\text{Grand total}}$$

For 'box' A: $E = \frac{150 \times 89}{300} = 44.5.$

For 'box' B: $E = \frac{150 \times 100}{300} = 50.$

For 'box' C: $E = \frac{150 \times 111}{300} = 55.5.$

For 'box' D: $E = \frac{150 \times 89}{300} = 44.5.$

For 'box' E: $E = \frac{150 \times 100}{300} = 50.$

For 'box' F: $E = \frac{150 \times 111}{300} = 55.5.$

STEP 4 The degrees of freedom are calculated by the formula:
df = (number of rows − 1) × (number of columns − 1)
= (2 − 1) × (3 − 1)
= 2

STEP 5 Since df = 2, the formula used to work out chi is:

$$\chi^2 = \sum \frac{(O-E)^2}{E}$$

Follow the working in Table 16.

Table 16

Box	Observed frequency	Expected frequency	O−E	$(O-E)^2$	$\frac{(O-E)^2}{E}$
A	80	44.5	35.5	1260.25	28.32
B	20	50	−30.0	900.00	18.00
C	50	55.5	− 5.5	30.25	0.55
D	9	44.5	−35.5	1260.25	28.32
E	80	50	30.0	900.00	18.00
F	61	55.5	5.5	30.25	0.55

Total = 93.74

χ^2 = the total of the final column = 93.74

The table of chi values (Table 17) shows that with two degrees of freedom the probability of obtaining a chi squared value of 93.74 is less than 1%. (In fact our value is way above the value of 9.21 that we need to find for significance at this level; values as high as this are only found when there is a marked difference in the observed frequencies of different categories, as in this example where the differences in choices of construction and textile industries is particularly marked.)

This means that the distribution of choices is unlikely to have occurred by chance, so we can accept the hypothesis that there is an association between sex and preference for working in the construction, textile or banking industries. The table of results shows us what exactly this preference was. This is a very obvious example since the ratio of males to females in each case is striking. Many such analyses are not as clear-cut, however, and care must be taken in interpreting the table to decide where the association between categories lies.

Table 17 CHI SQUARED TABLES

df	*p=10%*	*p=5%*	*p=1%*	*df*	*p=10%*	*p=5%*	*p=1%*
1	2.706	3.841	6.635	16	23.542	26.296	32.000
2	4.605	5.991	9.210	17	24.769	27.587	33.409
3	6.251	7.815	11.345	18	25.989	28.869	34.805
4	7.779	9.488	13.277	19	27.204	30.144	36.191
5	9.236	11.070	15.086	20	28.412	31.410	37.566
6	10.645	12.592	16.812	21	29.615	32.671	38.932
7	12.017	14.067	18.475	22	30.813	33.924	40.289
8	13.362	15.507	20.090	23	32.007	35.172	41.638
9	14.684	16.919	21.666	24	33.196	36.415	42.980
10	15.987	18.307	23.209	25	34.382	37.652	44.314
11	17.275	19.675	24.725	26	35.563	38.885	45.642
12	18.549	21.026	26.217	27	36.741	40.113	46.963
13	19.812	22.362	27.688	28	37.916	41.337	48.278
14	21.064	23.685	29.141	29	39.087	42.557	49.588
15	22.307	24.996	30.578	30	40.256	43.773	50.892

Source: Adapted from R. A. Fisher, *Statistical Methods for Research Workers* (Oliver and Boyd, 1938).

Using the chi squared tables

Look across the row which starts with your df value. If your chi value is lower than any of the values shown, then the probability that your result was due to chance (p) is greater than 10%. Find the number in the table which is closest to but smaller than your chi value. The probability that your result was due to chance is less than the value at the top of the column in which this number occurs. If your chi value is equal to one of the numbers in the row, then p = the value shown at the top of the column in which the t value is shown.

All p values refer to a two-tailed test and should be halved if your experimental hypothesis was one-tailed.

Examples

If your df = 1 and chi squared = 2.706, then there is a 10% probability that chance factors were responsible for the difference between the observed results and those expected if there was no relationship between the variables under study. The result is not significant.

If your df = 1 and chi squared = 4.2, then there is less than a 5% probability that chance factors were responsible for the difference between the observed results and those expected if there was no relationship between the variables under study. The result is significant.

What is the probability that the following chi values are a result of chance factors? (The answers are at the foot of the page.)

(a) df = 10 chi squared = 15.987
(b) df = 5 chi squared = 17.00
(c) df = 15 chi squared = 29.13

Answers: (a) p = 10%; (b) p = less than 1%; (c) p = less than 5%.

Goodness of fit

Chi Squared can be used to find out whether your data conform to a particular theoretical distribution. For example, do they differ significantly from a normal distribution? When used in this way it is known as a chi squared test for goodness of fit. The only difference to the normal chi squared test is that the expected frequencies are from the theoretical distribution rather than being calculated from row and column totals on a 'box table'. A common use for goodness of fit is for finding out whether there is a particular bias in a choice of items or activities, given that if there is no bias each activity would be chosen an equal number of times.

Example

Here is an example of the use of chi squared for goodness of fit, in an investigation of the choices of soap powder made by customers in a large supermarket. If there was no bias in choice we would expect each brand to be chosen equally often. There were 5 different brands available and the investigator observed 100 customers, each of whom bought one type of soap powder.

STEP 1 Since there is only one type of variable (type of soap powder), we cannot produce the same sort of box table as used in our previous chi squared example. So we simply list the frequencies of choice of each brand.

	A	B	C	D	E
Number of times each brand was chosen	10	30	30	10	20

STEPS 2 AND 3 The expected number of choices (E) of each brand, if there is no particular preference, is worked out using the formula:

$$E = \frac{\text{Total number of observations}}{\text{Number of alternative brands}} = \frac{100}{5}$$

Thus if there were no preference, each brand would be chosen 20 times.

STEP 4 The df value is found using the formula:

$$df = \text{number of alternatives} - 1$$

Therefore df = 4.

STEP 5 Since df is greater than 1, the formula for chi squared is:

$$\chi^2 = \sum \frac{(O-E)^2}{E}$$

The table below shows how this is calculated.

Brand	Observed frequency	Expected frequency	O−E	$(O-E)^2$	$\frac{(O-E)^2}{E}$
A	10	20	−10	100	5
B	30	20	10	100	5
C	30	20	10	100	5
D	10	20	−10	100	5
E	20	20	0	0	0
				Total	20

Chi squared = 20

The chi squared tables on page 50 show that there is a less than 1% probability of obtaining a chi squared value of 20 with a df value of 4, therefore the results do not fit the distribution expected if there were no bias in choice of soap powder. Brands C and B are significantly preferred.

Related t test

Experimental design: Repeated measures or matched pairs.
Minimum level of data: Interval/ratio.
What the test does: The test looks at the actual size of the differences between each subject's scores in the two conditions. To determine the likelihood that the difference between the two conditions was due to chance, the test takes into account the mean difference for the group as a whole and the variation in the size of the differences for individual subjects. It is unlikely that you would find a large mean difference between scores in the two conditions combined with little variety in the difference scores for each subject unless there was a real difference between the conditions.

(1) For the pair of scores from each subject:
(a) Subtract the score in condition B from that in condition A, to produce the subject's d score.
(b) Square each d score.

(2) Add up all the d scores. (Σd)

(3) Add up all the d^2 scores. (Σd^2)

(4) Divide Σd by n (the number of subjects) to find the mean d score. $(\bar{d})$

(5) Find t using the formula below:

$$t = \frac{\bar{d}}{\sqrt{\frac{\Sigma d^2 - (\Sigma d)^2/n}{n(n-1)}}}$$

(6) Use the t table on page 60 (Table 20) to find out whether the result is significant. To do this you will need the t score and the degrees of freedom (df). The value of df is found using the formula:

$$df = n - 1$$

Example

Here is an example of the use of a related t test, applied to the results of an experiment to test the hypothesis that there is a significant difference between scores on a test performed in quiet or noisy conditions.

Table 18 RESULTS

SUBJECT	SCORES IN CONDITION A (QUIET)	SCORES IN CONDITION B (NOISY)	STEP 1A A MINUS B d SCORES	STEP 1B d^2
1	20	17	3	9
2	18	17	1	1
3	15	13	2	4
4	14	14	0	0
5	14	12	2	4
6	12	13	–1	1
7	10	8	2	4
8	9	8	1	1
9	9	7	2	4
10	7	4	3	9
		Totals	$\Sigma d=15$	$\Sigma d^2=37$

STEP 1a The scores in condition B are subtracted from those in A, producing the d scores in Table 18.

STEP 1b Each d score is squared. See Table 18.

STEP 2 The column of d scores is added up. (Note that the -1 in the column reduces the size of this total.) $\Sigma d=15$

STEP 3 The column of d^2 scores is added up. $\Sigma d^2=37$

STEP 4 The mean d score is found by dividing Σd by n (the number of scores). $\Sigma d \div n = 15 \div 10$ $\bar{d}=1.5$

STEP 5 The value of t is found using the formula:

$$t = \frac{\bar{d}}{\sqrt{\frac{\Sigma d^2 - (\Sigma d)^2/n}{n(n-1)}}} = \frac{1.5}{\sqrt{\frac{37-(15)^2/10}{10(10-1)}}}$$

$$= \frac{1.5}{\sqrt{\frac{37-22.5}{90}}} = \frac{1.5}{\sqrt{0.1611}}$$

$$t = 3.7372$$

STEP 6 Df = n−1, therefore df = 9. The table on page 60 shows that for df = 9 and t = 3.7372, the probability that the result was due to chance is less than 1%. The result is therefore significant; we can reject the null hypothesis and accept that performance on this type of test is impaired by noise.

Unrelated t test

Experimental design: Independent subjects.
Minimum level of data: Interval/ratio.
What the test does: The test determines the probability that the difference between the two conditions is significant by inspecting the amount of difference between the two means and taking into account the variation of scores in the two conditions.

(1) Square each subject's score.

(2) Add up the scores in condition A. (ΣA)

(3) Add up the squares of the scores in condition A. (ΣA^2)

(4) Add up the scores in condition B. (ΣB)

(5) Add up the squares of the scores in condition B. (ΣB^2)

(6) Find the mean of the scores in condition A by dividing the total score by Na (the number of subjects in condition A). ($\bar{A}$)

(7) Find the mean of the scores in condition B by dividing the total score by Nb (the number of subjects in condition B). ($\bar{B}$)

(8) Find t using the following formula:

$$t = \frac{\bar{A} - \bar{B}}{\sqrt{\left(\frac{[\Sigma A^2 - (\Sigma A)^2/Na] + [\Sigma B^2 - (\Sigma B)^2/Nb]}{(Na-1)+(Nb-1)}\right) \times \left(\frac{1}{Na} + \frac{1}{Nb}\right)}}$$

This looks daunting but all you have to do is substitute the values that you have found in steps 1 to 7 into the formula. The example gives more help on working out the formula.

(9) Use the t table on page 60 to find out whether the result is significant. To do this you need t and the degrees of freedom (df). The df value is calculated using the formula:

$$df = (Na - 1) + (Nb - 1)$$

Example

Here is an example of the use of an unrelated t test, applied to the results of an experiment to test for sex differences in the short-term memory span of 12-year-old girls and boys when recalling lists of letters. The results given in Table 19 show the maximum number of letters perfectly recalled immediately after presentation.

Table 19

RESULTS

	SCORES IN CONDITION A (GIRLS)	STEP 1 A^2	SCORES IN CONDITION B (BOYS)	STEP 1 B^2
	7	49	6	36
	9	81	8	64
	7	49	8	64
	7	49	6	36
	6	36	4	16
	9	81	7	49
	5	25	5	25
	5	25	4	16
	5	25	5	25
	4	16	4	16
			5	25
Totals	$\Sigma A=64$	$\Sigma A^2=436$	$\Sigma B=62$	$\Sigma B^2=372$

STEP 1 Each subject's score is squared. See Table 19.

STEPS 2, 3, 4 and 5 The totals for each of the columns in Table 19 are found. See the table.

$\Sigma A=64$
$\Sigma A^2=436$
$\Sigma B=62$
$\Sigma B^2=372$

STEP 6 There are 10 subjects in condition A so the mean of the scores in A is 64 ÷ 10. $\bar{A}=6.4$

STEP 7 There are 11 subjects in condition B so the mean of the B scores is 62 ÷ 11. $\bar{B}=5.6364$

STEP 8 The value of t is found using the formula below.

$$t = \frac{\bar{A}-\bar{B}}{\sqrt{\left(\frac{[\Sigma A^2-(\Sigma A)^2/Na]+[\Sigma B^2-(\Sigma B)^2/Nb]}{(Na-1)+(Nb-1)}\right)\times\left(\frac{1}{Na}+\frac{1}{Nb}\right)}}$$

The values found in steps 1 to 7 are now substituted.

$$t = \frac{6.4-5.6364}{\sqrt{\left(\frac{[436-(64)^2/10]+[372-(62)^2/11]}{(10-1)+(11-1)}\right)\times\left(\frac{1}{10}+\frac{1}{11}\right)}}$$

First we work out everything inside brackets.

$$t = \frac{6.4-5.6364}{\sqrt{\left(\frac{[26.4]+[22.5455]}{(9)+(10)}\right)\times(0.1909)}}$$

$$= \frac{0.7636}{\sqrt{\left(\frac{48.9455}{19}\right)\times(0.1909)}} = \frac{0.7636}{\sqrt{2.5761\times 0.1909}}$$

$$= \frac{0.7636}{\sqrt{0.4918}} = \frac{0.7636}{0.7013}$$

$$t = 1.0888$$

The df value (Na−1) + (Nb−1) is 19. Table 20 on page 60 shows us that there is a greater than 10% probability of getting a t value of 1.0888 by chance when df=19. The difference between the results of the girls and boys is therefore not significant. The null hypothesis is accepted, i.e. there is no real difference between twelve-year-old girls and boys in terms of short-term memory span for letters.

Table 20 FOR USE AFTER RELATED OR UNRELATED t TEST

df	*p=10%*	*p=5%*	*p=1%*	*df*	*p=10%*	*p=5%*	*p=1%*
1	6.314	12.706	63.657	16	1.746	2.120	2.921
2	2.920	4.303	9.925	17	1.740	2.110	2.898
3	2.353	3.182	5.841	18	1.734	2.101	2.878
4	2.132	2.776	4.604	19	1.729	2.093	2.861
5	2.015	2.571	4.032	20	1.725	2.086	2.845
6	1.943	2.447	3.707	21	1.721	2.080	2.831
7	1.895	2.365	3.499	22	1.717	2.074	2.819
8	1.860	2.306	3.355	23	1.714	2.069	2.807
9	1.833	2.262	3.250	24	1.711	2.064	2.797
10	1.812	2.228	3.169	25	1.708	2.060	2.787
11	1.796	2.201	3.106	26	1.706	2.056	2.779
12	1.782	2.179	3.055	27	1.703	2.052	2.771
13	1.771	2.160	3.012	28	1.701	2.048	2.763
14	1.761	2.145	2.977	29	1.699	2.045	2.756
15	1.753	2.131	2.947	30	1.697	2.042	2.750

Source: Adapted from R. A. Fisher, *Statistical Methods for Research Workers* (Oliver and Boyd, 1938).

Using Table 20

Look across the row which starts with your df value. If your t value is lower than any of the values shown, then the probability that your result was due to chance (p) is greater than 10%. Find the number in the table which is closest to but smaller than your t value. The probability that your result was due to chance is less than the value at the top of the column in which this number occurs. If your t value is equal to one of the numbers in the row, then p = the value shown at the top of the column in which the t value is shown.

All p values refer to a two-tailed test and should be halved if your experimental hypothesis was one-tailed.

Examples

If your df = 1 and t = 6.314, then there is a 10% probability that the difference between the results of the two conditions was due to chance. The result is not significant.

If your df = 1 and t = 13.00, then there is less than a 5% probability that the difference between the results of the two conditions was due to chance. The result is significant.

Spearman's rho test

(This is a test of correlation which describes the relationship between two variables, X and Y. It does *not* test the difference between two sets of scores.)

(1) Rank the scores on variable X, giving the lowest score a rank of 1.

(2) Rank the scores on variable Y, giving the lowest score a rank of 1.

(3) Work out the difference between the two rankings for each subject. (d)

(4) Square this difference for each subject. (d^2)

(5) Add up the d^2 scores. (Σd^2)

(6) Use the following formula to work out the correlation co-efficient (r_s):

$$r_s = 1 - \frac{6(\Sigma d^2)}{N(N^2 - 1)} \qquad N = \text{number of subjects}$$

This gives the correlation co-efficient, which describes the relationship between the two variables. Use Table 21 to find out whether the correlation found is significant. If the co-efficient turns out to be significant, it means that it is unlikely to have been the result of chance and is likely to be found again.

The correlation is significant at the 5% level if it is equal to or more than the critical value in Table 21 for the number of pairs of scores. Notice that the critical value is lower if the direction of the correlation (positive or negative) was predicted before the study (one-tailed hypothesis).

Table 21

N	CRITICAL VALUE one-tailed	CRITICAL VALUE two-tailed	N	CRITICAL VALUE one-tailed	CRITICAL VALUE two-tailed
5	0.90	1.00	18	0.40	0.47
6	0.83	0.89	19	0.39	0.46
7	0.71	0.79	20	0.38	0.45
8	0.64	0.74	21	0.37	0.44
9	0.60	0.70	22	0.36	0.43
10	0.56	0.65	23	0.35	0.42
11	0.54	0.62	24	0.34	0.41
12	0.50	0.59	25	0.34	0.40
13	0.48	0.56	26	0.33	0.39
14	0.46	0.54	27	0.32	0.38
15	0.44	0.52	28	0.32	0.38
16	0.43	0.50	29	0.31	0.37
17	0.41	0.49	30	0.31	0.36

Source: J. H. Zar, 'Significance testing of the Spearman Rank Correlation Coefficient', *Journal of the American Statistical Association*, 1972, vol. 67.

Example

Here is an example of the use of Spearman's rho, to find the correlation between scores on a test taken by subjects in March and their scores on the same test in April. The prediction made before the study was that there would be a positive correlation between the two sets of scores.

STEP 1 The March scores are ranked (see Table 22). Refer to page 110 if you have difficulties with ranking.

STEP 2 The April scores are ranked. (See Table 22).

STEP 3 The difference between the March and April rankings of each subject is found (don't worry about whether the March or April ranking is biggest, we are only interested in the size of the difference). See Table 22.

STEP 4 The d score of each subject is squared. See Table 22.

Table 22

RESULTS

Subject	*March score*	*April score*	*Step 1 March ranks*	*Step 2 April ranks*	*Step 3 d*	*Step 4 d^2*
1	20	21	10	9	1	1
2	9	7	1	1	0	0
3	14	19	5.5	7.5	2	4
4	13	10	4	2	2	4
5	10	13	2	3	1	1
6	14	17	5.5	6	0.5	0.25
7	18	22	9	10	1	1
8	15	19	7	7.5	0.5	0.25
9	11	14	3	4	1	1
10	17	15	8	5	3	9
				Step 5	Total =	21.5

STEP 5 The total of the d^2 scores in the final column of the table above $(\Sigma d^2) = 21.5$.

STEP 6 The number of subjects, N, = 10.

$$r = 1 - \frac{6(\Sigma d^2)}{N(N^2 - 1)}$$

$$= 1 - \frac{6 \times 21.5}{10(100 - 1)}$$

$$= 1 - 0.13$$

$$r = +0.87$$

The table of critical values (Table 21) shows that for 10 pairs of scores, the minimum value for a correlation to be significant is 0.56 for a one-tailed hypothesis. The correlation of +0.87 found between the scores of subjects on this test in March and April is therefore significant.

Pearson's product moment correlation

This is a test of correlation which describes the relationship between two variables, X and Y. It is the parametric equivalent of Spearman's rho. (For a discussion of the meaning of parametric tests, see page 75.) Before calculating the co-efficient, draw a scattergram as it will give a good estimate of the size and direction of the correlation.

(1) Find the mean of the X scores. $\bar{X}$

(2) Subtract $\bar{X}$ from each X score to give a series of Dx scores. Dx

(3) Square each Dx score to give a series of Dx^2 scores. Add up the Dx^2 scores. ΣDx^2

(4) Find the mean of the Y scores. $\bar{Y}$

(5) Subtract $\bar{Y}$ from each Y score to give a series of Dy scores. Dy

(6) Square each Dy score to give a series of Dy^2 scores. Add up the Dy^2 scores. ΣDy^2

(7) For each pair multiply the Dx score by the Dy score to give a series of DxDy scores. Add up the DxDy scores. $\Sigma DxDy$

(8) Find the correlation co-efficient (r) using the following formula:

$$r = \frac{\Sigma DxDy}{\sqrt{(\Sigma Dx^2)(\Sigma Dy^2)}}$$

(9) Use the table on page 62 (Table 21) to find out whether the correlation is significant. (*Note:* The critical values are significant at slightly less than 5% for a Pearson's co-efficient.)

Example

Here is an example of the use of Pearson's test, to find the correlation between scores on a test taken by subjects in March and their scores on the same test in April. The prediction made before the study was that there would be a positive correlation between the two sets of scores. The data is the same as that we used in the Spearman's rho worked example.

Table 23

RESULTS AND CALCULATIONS

Subject	*March score*	*April score*	*Dx score*	*Dx^2 score*	*Dy score*	*Dy^2 score*	*DxDy score*
1	20	21	5.9	34.81	5.3	28.09	31.27
2	9	7	–5.1	26.01	–8.7	75.69	44.37
3	14	19	–0.1	0.01	3.3	10.89	–0.33
4	13	10	–1.1	1.21	–5.7	32.49	6.27
5	10	13	–4.1	16.81	–2.7	7.29	11.07
6	14	17	–0.1	0.01	1.3	1.69	–0.13
7	18	22	3.9	15.21	6.3	39.69	24.57
8	15	19	0.9	0.81	3.3	10.89	2.97
9	11	14	–3.1	9.61	–1.7	2.89	5.27
10	17	15	2.9	8.41	–0.7	0.49	–2.03
	Totals			$\Sigma Dx^2 = 112.9$		$\Sigma Dy^2 = 210.1$	*123.3

*$\Sigma DxDy$=

STEP 1 The mean of the March scores is 14.1.

STEP 2 The Dx scores are calculated by subtracting 14.1 from each of the March scores.
See Table 23.

STEP 3 Each Dx score is squared and the squares are added together.
See Table 23. $\Sigma Dx^2 = 112.9$

STEP 4 The mean of the April scores is 15.7.

STEP 5 The Dy scores are calculated by subtracting 15.7 from each of the April scores.
See Table 23.

STEP 6 Each Dy score is squared and the squares are added together. See Table 23. $\Sigma Dy^2 = 210.1$

STEP 7 The Dx and Dy scores are multiplied, giving the DxDy column in Table 23. This column is totalled to give ΣDxDy. $\Sigma DxDy = 123.3$

STEP 8 The correlation co-efficient (r) is found using the formula:

$$r = \frac{\Sigma DxDy}{\sqrt{(\Sigma Dx^2)(\Sigma Dy^2)}}$$

$$= \frac{123.3}{\sqrt{112.9 \times 210.1}}$$

$$r = +0.8$$

The table on page 62 shows that for 10 pairs of scores any correlation equal to or above 0.53 is significant at the 5% level. Our correlation of +0.8 is, therefore, significant.

Note that using the data in the example above, Pearson's test results in a correlation of +0.8 but the Spearman's correlation was +0.87. The two tests usually give slightly different results because they treat the data in a different way. Pearson's uses the information at the interval/ratio level while the Spearman test converts to the ordinal level. The authors prefer to use Spearman's rho because it is simpler to calculate.

Standard deviation

The formula for standard deviation (SD) is:

$$SD = \sqrt{\frac{\Sigma(X-\bar{X})^2}{N-1}}$$

where X is each score,
N is the number of scores, and
$\bar{X}$ is the mean of the scores.

Example

Here is an example of the use of the formula, to calculate the standard deviation of the following set of scores: 5, 7, 4, 8, 6.

STEP 1 The mean of the scores is found by dividing the total of the scores (30) by the number of scores (5). $\bar{X} = 6$, $N = 5$

STEP 2 $X - \bar{X}$ is found by subtracting the mean from each score. See the table below.

STEP 3 $(X - \bar{X})^2$ is then calculated. See the Table below.

X (Raw score)	$X-\bar{X}$ (Step 2)	$(X-\bar{X})^2$ (Step 3)
5	−1	1
7	1	1
4	−2	4
8	2	4
6	0	0
Total 30		Total 10

STEP 4 $\Sigma(X-\bar{X})^2$ is found by adding up the numbers in the $(X-\bar{X})^2$ column. $\Sigma(X-\bar{X})^2 = 10$

STEP 5 The standard deviation is found by substituting into the formula:

$$SD = \sqrt{\frac{\Sigma(X-\bar{X})^2}{N-1}} = \sqrt{\frac{10}{4}} = \sqrt{2.5}$$

$$SD = 1.58$$

Table 24

TYPES OF DATA AND THEIR ANALYSIS

	Counting NOMINAL DATA	*Ranking* ORDINAL DATA	*Scaling* INTERVAL/ RATIO DATA
EXAMPLES	Like 5 Dislike 7	1	lbs, oz, ft, ins, cms, gms
	Men 9 Women 12	2	
	Yes 14 No 1	3	
AMOUNT OF INFORMATION IN THE DATA	Little	More	Most
MEASURE OF AVERAGE	Mode	Median	Mean
MEASURE OF VARIATION		Interquartile range	Standard deviation
TESTS FOR CORRELATION		Spearman	Pearson
TESTS OF SIGNIFICANT DIFFERENCE	Chi squared Sign test	Wilcoxon Mann-Whitney	Related t test Unrelated t test

Significance

Statistical tests such as the sign test, Mann-Whitney, Wilcoxon, chi squared and the t tests tell us how likely it is that any difference found in a dependent variable under two conditions was due purely to chance factors. If a result occurred due to chance, we cannot rely on it.

For example, a man tossed a coin 100 times and obtained 70 heads. A woman tossed the same coin 100 times and obtained 40 heads. It is quite likely that, if we repeated this experiment a few times, sometimes the man would get more heads, sometimes the woman; the result is not reliable since the difference between the results of the man and the woman are due to chance factors over which we have no control and which vary at random.

A man tossed a two-headed penny 100 times and obtained 100 heads, then threw a two-tailed penny and obtained no heads; this result is not due to chance and will be found again if the demonstration is repeated.

In most experiments it is not so clear whether the result occurred due to the independent or to the chance variables – statistical tests tell us the probability that the result occurred due to chance.

Probability

Probability can be expressed in terms of percentages – if an event is 100% probable, then it is certain to happen; if it is 0% probable, it is certain that it will not happen. In practice, these extreme probabilities are never found.

Probability can also be expressed on a scale ranging from 0 to 1. Probability of 1 is equivalent to 100%. Probability of 0 is equivalent to 0%.

At the end of a statistical test of significance you look at the relevant table which tells you the value of p.

Table 25

Probability in percentage	*Equivalent probability on the 0 to 1 scale*
100%	0.1
50%	0.5
10%	0.1
5%	0.05
1%	0.01

p = *the probability that the result occurred due to chance factors*

If p is large, then the result is unreliable since the difference between the results of the 2 conditions was probably due to chance factors. If condition A was better than condition B, a repeat of the experiment might result in B being better than A.

If p is small, then the difference between the two sets of results is unlikely to be due to chance – it is therefore likely to be due to the independent variable and should be repeatable.

A very important question, therefore, is, at what value can we call p small?

There is no absolute answer to this but the most common **level of significance** is 5%. This means that if p is 5% (0.05) or smaller, the result is said to be **significant**, that is, the result is unlikely to have occurred by chance and is likely to be repeatable. If p is greater than 5%, the result is said to be **not significant.**

Table 26

EXAMPLES

These values of p are significant (using the 5% level of significance)				
5%	4%	3%	2%	1%
0.05	0.04	0.03	0.02	0.01

A value of p = 1% is less likely to have occurred by chance than a value of p = 5%.

If a result is significant we can conclude that the difference between the results of the two conditions was probably due to the independent variable.

Table 27

EXAMPLES

These values of p are not significant				
70%	50%	30%	10%	6%
0.7	0.5	0.3	0.1	0.06

Since significance is a matter of probability rather than certainty, we can sometimes be wrong in assuming that a significant difference was due to the independent variable or that a not significant result was due to chance factors. These mistakes are known, respectively, as **Type 1** and **Type 2** errors.

Type 1 error

When saying that a result is significant we may be making an error, since there is still a small possibility that the result was due to chance. If we say a result is significant but it turns out to be an unreliable result, we have made a Type 1 error.

Type 2 error

If p is more than 5% the result is not significant. However there is a possibility that the result was not due to chance but due to the independent variable. If we say a result is not significant but it turns out that the results are genuine, a Type 2 error has occurred.

Probability of error

Remember that statistics deal with probabilities, not certainties. It is less likely that you have made a Type 1 error if $p = 1\%$ than if $p = 5\%$.

It is less likely that you have made a Type 2 error if $p = 20\%$ than if $p = 6\%$.

A Type 2 error is most likely to occur if you only have a few subjects, since the tests need a lot of information to give a p value of less than 5%.

If you do a statistics test and find $p = 5\%$, this means that the result is significant. The probability that the result occurred due to chance is only 5% and since this is not very high the difference between the results under the two conditions was probably due to the independent variable. In some circumstances where there is a high cost involved in making a Type 1 error, the level of significance is set at 1%, meaning that the result is only regarded as significant if the probability that it was due to chance is 1% or less. Obviously when this is done the chances of making a Type 2 error increase correspondingly. There is no absolute rule about which level of significance to choose. It is up to the researcher, taking into account the effect of Type 1 and Type 2 errors. If you were testing a new drug which by the nature of things might have unknown side effects, you would want to be very sure that it really did do what it was meant to do and so would use a 1% level of significance, but in most circumstances the 5% level gives a reasonable balance between the possibilities of the two types of error.

Hypotheses

Experiments and statistics test hypotheses (ideas) about the effect of certain variables on others. There are two main types of hypothesis. Both are present at the beginning of an experiment – the result and the decision about significance tells you which one is most likely to be true.

Experimental hypothesis

The **experimental hypothesis** states that there will be a significant difference between conditions. (This may be a two-tailed experimental hypothesis where the direction of the difference is not specified, or a one-tailed experimental hypothesis which predicts the direction of the difference, for instance that condition A will be faster than condition B.) Whether the hypothesis is one- or two-tailed, it should be made in unambiguous terms. The hypothesis ‘bulls will be angrier when presented with a red rag than when presented with a blue rag', for example, is not satisfactory because we don't know how anger is to be measured; anger needs to be operationally defined so that it is clear what we mean. 'Bulls will charge an experimenter more often during a one-minute period if he is waving a red flag than if he waves a blue flag' is a better hypothesis, though you might want to say more about the sort of bull and the speed of flag waving. Whether the hypothesis is true or not is a matter for experimentation.

Null hypothesis

The **null hypothesis** states that there will be no significant difference between the conditions. By deciding the probability that the difference between the results of the two conditions was due to chance, the statistical test is commenting on the likelihood of the null hypothesis being true.

Note that statistical tests test only the null hypothesis; when they give the probability that the difference between the results of the conditions was due to chance they are reporting the probability of the null hypothesis being true.

Accepting hypotheses

The experimental hypothesis is accepted as a result of elimination of the other possible explanations for the difference between the results of the two or more conditions. A difference between the results could be due to:

(a) confounding variables – these are eliminated by good experimental design with adequate control procedures.

(b) chance – this is eliminated by the use of statistical tests.

(c) independent variables – we accept that these caused the difference only if we are reasonably sure that (a) and (b) did not.

If the result is significant, you can reject the null hypothesis and accept the experimental hypothesis.

If the result is not significant, you can reject the experimental hypothesis and accept the null hypothesis.

The choice of table to use after a statistical test to determine the probability that the null hypothesis is true depends on whether the experimental hypothesis was one- or two-tailed. All the tables in this book are two-tailed tables, to be used in the normal way when a two-tailed experimental hypothesis has been set. These tables can, however, be used for one-tailed hypotheses if we realise that the reported probabilities of the result being due to chance can be halved if we have a one-tailed hypothesis. This is because the two-tailed tables give you the combined probabilities of the difference between the conditions being a result of condition A being greater than B and B being greater than A; for a one-tailed hypothesis you have already excluded one of these and so if the table says 10% for a two-tailed test you can treat this as 5% for a one-tailed test. (A one-tailed test is simply a statistical test done after proposing a one-tailed hypothesis. The only difference to a two-tailed test is the way the tables are used after working out the test.) If you now feel tempted to state one-tailed hypotheses all the time, don't, because if the result turns out to be in the opposite direction than you predicted then no difference, no matter how big, can be taken as significant. Only use one-tailed hypotheses if theory and previous observations lead you to be *very* sure of it.

Parametric and non-parametric tests

There are basically two different types of statistical test.

Non-parametric tests: These tests can be recognised by the fact that they use **ranked** or **nominal data**. Examples include the Mann-Whitney, Wilcoxon and Spearman's rho. These tests are relatively new compared with the older parametric tests.

Parametric tests: These can be recognised because they use **interval** or **ratio data**; they do not rank data, they use it as it is. Examples of these tests are the related t test and the unrelated t test. The data must fulfil a set of assumptions made by parametric tests (see below).

The experimental design and the type of data you collect determine which test you use. If you have interval or ratio data, you may be able to use either parametric or non-parametric tests.

Table 28

EXPERIMENTAL DESIGN	NON-PARAMETRIC TEST	PARAMETRIC TEST
Repeated measures	Wilcoxon	Related t test
Independent subjects	Mann-Whitney	Unrelated t test

Parametric statistics make assumptions about the data that cannot always be fulfilled in psychological investigations.

ASSUMPTION 1

Parametric tests assume that the scores you have obtained in both conditions are drawn from a **normally distributed** population of scores (as opposed to a skewed or other type of distribution).

ASSUMPTION 2

Parametric tests also assume that the two normal distributions have the same **variance** (the standard deviation squared). See pages 14 and 88.

ASSUMPTION 3

Parametric tests assume that the results of the experiment are **accurate** as **interval** or **ratio data**. There is an argument that many psychological measures, such as intelligence or personality, cannot be measured accurately at the ratio or interval levels.

If your data does not fulfil these criteria, the use of a parametric test may give you an incorrect conclusion. (The t tests, however, are an exception, and are said to be **robust**, that is, they can cope with data which does not fully meet the assumptions made by parametric tests in general. The independent t used with different numbers in each group, however, is not robust.)

An advantage of parametric tests

Since parametric tests use interval or ratio data, they have more information available to them than the non-parametric tests. Therefore, parametric tests are more likely to find a significant result than the non-parametric tests if the difference between the conditions was due to the independent variable rather than to chance factors. Parametric tests are, therefore, said to be **more powerful** or **sensitive**.

Many psychologists assume that their data fulfils the parametric assumptions and, especially when using t tests, simply take a cursory glance to see whether the data is obviously breaking the rules. It is possible to test that the data is from normally distributed populations (using a chi squared test for goodness of fit) and that there is equal variance (using an F test), but most people working on a student practical won't spend time on this. Sidney Siegel, in his book *Nonparametric Statistics for the Behavioural Sciences*, argues powerfully that our subject matter and methods do not produce the sort of data required by the parametric tests and that we should abandon them; such is the reinforcing power of a few extra significant results that his message has been largely ignored.

Assumptions made by both parametric (t test, etc.) and non-parametric (Wilcoxon, etc.) statistics

They both assume that members of the sample are representative of whatever population is being studied and that the data is accurate in the form used by the test. No test can help you reach the right conclusion unless the experiment was well designed and controls all possible confounding variables.

Reliability and validity

Imagine that we had produced a new test which measured people's reading ability but each time you took the test it gave you a different score, or you gained one score when it was marked by Fred Bloggs and a different one when marked by Joe Smith. The test would, quite rightly, be criticised on the grounds that it was not **reliable**. Reliability refers to the consistency of a measuring instrument. Things like rulers and stop watches are usually very reliable; it doesn't matter who uses them, they give the same answer each time a particular line or length of time is measured. Psychological tests cannot usually achieve such a high degree of reliability but a reasonable level should be reached or our results will be worthless.

Reliability alone is not enough. A test of intelligence which simply involved measuring the distance between the subject's eyes would presumably be very reliable, but not many people would accept it as a good test of intelligence because there is no evidence that it really does test intelligence; the test would not be **valid**. Validity is a matter of whether the measuring implement measures what it is supposed to.

The reliability and validity of psychological tests can be investigated using correlational techniques.

Methods of testing reliability

Scorer reliability

This looks at the level of agreement between different individuals marking the performance of the same group of subjects. The scores obtained by subjects when marked by one trained scorer are correlated with those obtained with the other trained scorer. If the correlation between the two sets of scores is high, the test is said to have scorer reliability. Closed-ended multiple choice type questions have great scorer reliability; open-ended essay type questions are the most suspect as far as this type of reliability is concerned, and it is usually necessary to train the markers of this type of test. Always check scorer reliability first, even if you only do this by looking at the type of question asked (open or closed). Then use one of the following methods for finding out whether the test itself is reliable.

Test-retest reliability

Subjects take the test twice and if the test is reliable the two scores should be highly correlated. The second test must be taken after a period of time so that the subject does not simply recall what he did last time. If the length of time is too great, however, the subject may have changed in some way and a low correlation may simply reflect this. The period between test and retest is usually a few weeks, though for some psychological characteristics (e.g. motivation) this is much too long to expect the subject to remain the same.

Equivalent forms

This method avoids the problems of the time period involved in test-retesting, but two equivalent or similar forms of the same test are required. The two forms should be sufficiently dissimilar that the subject does not simply repeat himself when taking the second test. The time and effort required to produce such an equivalent form is usually considerable. The degree of correlation between subjects' scores on each test reflects the reliability of the test, assuming that the tests really are equivalent.

Split half

This is more a measure of internal consistency (i.e. the extent to which each part of the test gives the same results as other parts of the test), but useful if it isn't possible to repeat the test and there is no equivalent form. After the test has been taken, subjects' performance on one half of the test is correlated with their performance on the other half (in effect the halves are used as equivalent forms). If the test had one hundred questions, it would be best to take all the even-numbered questions as one half and the odd-numbered questions as the other. This avoids the possibility that one half of the test is very different from the other, as might occur with tests such as those of IQ where the early questions are often easier than later ones.

Methods of testing validity

Content or face validity

The easiest way to discover whether a test is valid is to examine it and decide whether it *looks* as though it is. If independent experts in the subject area agree that the test looks as though it does what it is supposed to do, then the test is said to have content validity. This is a very crude way of validating a test and other methods should also be used if possible.

Predictive validity

The results of many tests are used to predict future behaviour. For example, the interview and other procedures used in job selection are supposed to predict who would be best at the job if appointed. The predictive validity of a test is investigated by correlating the prediction made at the time of the test with later performance. For procedures such as job selection there is a practical difficulty in attempting a full-blown predictive validity investigation because we should really look at the correlation between test and performance of those who did poorly on the interview as well as those who succeeded; the only way to do this is to appoint all the people who come for interview during the validation study!

Concurrent validity

This method compares the test scores with another independent method of testing the same psychological variable. For example, a new test of reading ability might be validated by correlating children's marks on the test with estimates of the children's ability made by their teachers.

Construct validity

If a test of anxiety is valid, it should give a high reading in situations where our theories of human nature tell us that people are usually anxious. So we might try out the test on students who are just about to enter an examination room or parachutists just before their first jump. If the test results in these conditions were not considerably different to those found in more relaxed circumstances, we would question the test's validity. This method of validating a test, by comparing test results with what would be expected as a result of a common sense or 'academic' theory, is known as construct validity.

The greater the number of ways a test has been shown to be reliable and valid, the more confident we can be when using it in our research. Evidence of reliability and validity is usually published with any new test, together with instructions for standardised methods of administering and scoring the test and descriptions of normal test results for specific groups of people. Be very wary of any test which is not accompanied by this information.

Observational methods

There are many situations where behaviour is so complex that it cannot be isolated clearly enough to be manipulated experimentally. In addition, when researchers are starting to look at a type of behaviour not previously studied, observational studies must first be made before any hypothesis about the causes of such behaviour can be formed.

A major problem with observational studies is the lack of time which observers have to make a record of their observations. If the observer also has to record the type of behaviour seen, so much time has to be spent in writing that observations may be missed altogether. To overcome this problem, behavioural categories can be developed. Before the study begins the observer decides what types of behaviour are to be recorded and then draws up a table of categories which can simply be ticked when the particular behaviour occurs. Operational definitions of behaviour must be devised so that the observer has clear criteria for what constitutes a particular category of behaviour.

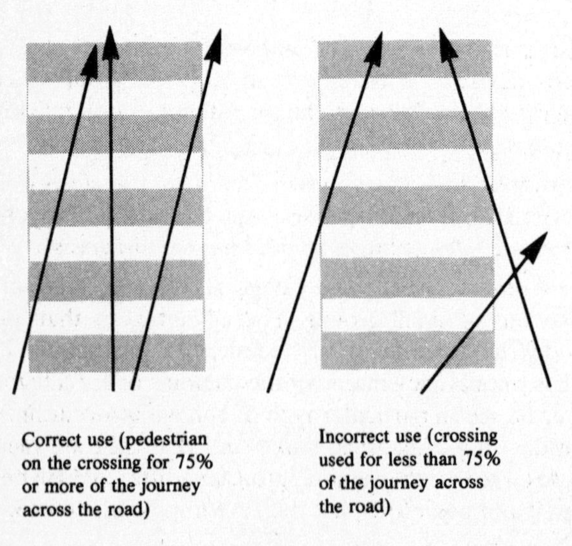

Figure 18

For example, if you were conducting an observational study of the behaviour of males and females on zebra crossings, you would need an operational definition of what constitutes 'correct' use of the crossing. If, for example, an individual walks across the road one metre to the side of the crossing, have they used the crossing correctly? You would need to define exactly what constitutes 'correct' and 'incorrect' methods of crossing, and agree this in advance so that all observers record the behaviour in the same way (see Figure 18).

Once the definitions are agreed, observers simply use ticks to record how many males and females used the crossing correctly and how many used it incorrectly. Figure 19 gives an example.

Male correct ///// /////

Male incorrect ///// ///// ///// /////

Female correct ///// ///// ///// ///// ///// /////

Female incorrect ///// //

Figure 19

Another example of this approach are the observational categories devised by Bales* when studying group discussions. Bales arrived at the following major categories of behaviour:

(a) Emotionally positive responses, e.g. agrees.
(b) Problem-solving responses – answers, e.g. gives opinion.
(c) Problem-solving responses – questions, e.g. asks opinion.
(d) Emotionally negative responses, e.g. disagrees.

Trained observers use these categories when they are observing the behaviour of small groups. The advantage is that instead of having to write down every tiny detail of behaviour, the observer simply has to put a tick in one of the columns of his category sheet whenever he sees a particular type of behaviour occuring.

When devising categories, make sure that they are meaningful and not so large that they ignore important differences in behaviour or so small and numerous that they are impossible to use.

*R. F. Bales, *Interaction Process Analysis: A Method for the Study of Small Groups* (Addison-Wesley, 1950).

In observational studies of animal behaviour, a specific method of recording observations called the **ethogram** has been developed. The ethogram is a detailed catalogue of the behaviour of an animal. This catalogue of behaviour is made up of two kinds of description. The first is the description of motor patterns, literally, the description of the physical movements made by the animal. This has the advantages that it is relatively objective, since it avoids the possibility of anthropomorphism (assigning human attributes to animals), and is accurate. However its disadvantages are that it does not relate the animal's movements to its environment, and that it can also be cumbersome to use – often all individual limb movements have to be described. The second kind of description is description by consequence. In this method the description is of consequences of the animal's behaviour in relation to its environment. As an example, consider the description 'The rat approached the lever'. If this activity were being described by motor pattern alone, the description would have been concerned with the limb movements and how the animal moved. Description by consequence is more concerned with why the animal moved – what the consequences were of the movement. The advantages of description of behaviour by consequence are that it is a more economical form of description, and that it relates the animal to its surroundings. But it has the disadvantage that it is not appropriate to the study of detailed differences in behaviour, or of motor patterns, between different animals. In order to obtain a full description of animal behaviour, researchers usually use both forms of description.

The example below is taken from Niko Tinbergen's work on territorial and courtship behaviour in the stickleback ('The curious behaviour of the stickleback', *Scientific American*, December 1952, vol. 187, no. 6, pp. 22–6). Description by consequence is enclosed in round parentheses, whilst description by motor pattern is enclosed in square parentheses.

> . . . (The threat display of male sticklebacks is of two types. When two males meet at the border of their territories, they begin a series of attacks and retreats. Each takes the offensive on his own territory, and the duel seesaws back and forth across the border. Neither fish touches the other, the two dart back and forth) as though attached by an invisible thread. . . .

When the fight grows in vigour, however, the seesaw manoeuvre may suddenly change into something quite different. [Each fish adopts an almost vertical head-down posture, turns its side to its opponent, raises its ventral spines and makes jerky movements with the whole body.]

Observational studies – A general planning sequence

(1) Having decided the behaviour(s) to be observed, draw up a checklist for observers, so as to make the recording of the behaviour easier and more reliable. The most common type of checklist is one which uses behaviour categories (see page 82).
(2) Establish the reliability of your observations. The most frequent method of doing this is to employ two or more observers who observe the behaviour independently using the same behaviour categories. A general rule of thumb is that if there is a 90–95% agreement between observers' recordings of the same events, you can assume your method is reliable.
(3) Estimate the minimum observation time needed for accurate recordings of the behaviour. This will vary according to the complexity and frequency of the behaviour, but, as a general guide, you should spend a minimum of 2 hours on your observations.
(4) You may not have time to oberve behaviour for 24 hours each day, or the behaviour you wish to observe may only occur at a particular time each day. Often therefore you can only make observations of samples of behaviour. This is known as *time sampling*. Two main methods are used to obtain representative samples of behaviour:

(a) The length of time over which the behaviour occurs is divided into 'time units', usually of a few minutes each. You then decide what proportion of the total time units you wish to observe and select the required number at random. For example, if the total time of 240 minutes were made up of 40 time units of 6 minutes each and you decided that you wanted to sample 30% of the total time, you would need to choose 30% of 40 = 12 time units, at random, to observe.
(b) As an alternative, you could record observations every (say) 3 or 5 or 10 minutes throughout an hour, and choose a different hour each day.

(5) As with experimental designs, it is often very useful to conduct a small-scale pilot study first, in order to sort out any 'bugs' in your techniques.

Probably the major advantage of observational studies over experimental methods is that you can observe your subjects performing naturally in their normal environments – particularly important, for example, in studies of children's behaviour (particularly with parents or other children) and for studies of animal behaviour. Do remember, however, that experiments can be carried out in the 'normal environment' rather than in the laboratory, though it is more difficult to control extraneous variables.

Quick reference section

The contents of this section are arranged in groups of related concepts. Many of the concepts are considered in greater detail in other parts of the book (see Index).

MEAN — The arithmetic average obtained by adding up all the scores and dividing by the number of scores.
Advantage: It uses all the scores.
Disadvantage: It is influenced by extreme scores.

MEDIAN — The value that has as many scores above it as it has below (when the scores are placed in order of size).
Advantage: Not as influenced by extreme scores as is the mean.
Disadvantage: Does not use the arithmetic values of all scores and thus cannot be used for further arithmetic calculation.

MODE — The value that occurs most frequently.
Advantage: Easy to spot.
Disadvantage: There may be more than one.

RANGE — The difference between the highest and lowest scores.

STANDARD DEVIATION — A measure of the distribution of scores around the mean. A large standard deviation means that there is a wide scatter of scores whereas a small standard deviation means that most scores are very close to the mean value.

INTERQUARTILE RANGE

When you have a set of data for which the median gives a better description than the mean, then it would not be appropriate to use standard deviation to describe the variation in the data. Under these circumstances the interquartile range is a useful measure of variation, allowing you to compare the sets of data. The interquartile range is not as affected by extreme scores as is the range. It is calculated by putting the scores into ascending order and subtracting the value that comes 25% of the way along the scores from that which comes 75% of the way along.

VARIANCE

Another measure of the distribution of data around the mean. Mathematically it is equal to the square of the standard deviation of a set of scores.

EXPERIMENT

A situation where one set of variables (the independent) is deliberately manipulated to find the effect on another set (the dependent), whilst controlling extraneous variables.

FIELD EXPERIMENT

An experiment performed in the 'real world' rather than the laboratory. It is usually more difficult to control extraneous variables under these conditions, but the results of such experiments are more likely to relate to everyday behaviour.

EXTRANEOUS VARIABLE

Anything, other than the independent variable, which might have an effect on the result of your experiment.

CONFOUNDING VARIABLE

An extraneous variable which has not been controlled and may therefore have had a different effect in each condition. If a confounding variable was present, it is not possible to say whether the difference found between the two conditions was due to the independent or to the confounding variable.

CONSTANT OR SYSTEMATIC ERRORS

These are the result of bad experimental design which has allowed a variable other than the independent variable to have a consistent, predictably different effect on one condition than on the other. The variable thus becomes confounding. Order effects (see below) are examples of the sort of thing that can cause systematic errors if left uncontrolled.

RANDOM ERRORS

These errors creep in due to factors which are not so predictable as those which cause systematic errors, and as a result it is not so easy to reduce their effects. It is simple enough to make sure that in a visual perception experiment we avoid the constant error of putting all the people wearing glasses into just one condition. But there are many other ways in which our subjects may vary over which we have little control, such as what happened to them on the way to the laboratory, what they had for breakfast, their social relationships, etc. These things vary randomly in a population. Random errors cannot be eliminated, though they can and should be reduced as much as possible by techniques such as randomly allocating subjects to conditions. Subjects are often studied in laboratories in order to reduce as far as possible random environmental variables that occur outside. However temperature fluctuations, light changes and a hundred other things can vary in a lab without the experimenter realising it; fortunately most of these factors will have no effect on the results, but some *might*. After performing an experiment, statistical tests of significance are used to give an estimate of the likelihood that any difference found in the results of the two conditions was due to random error (chance).

DEMAND CHARACTERISTICS

Subjects are not passive recipients of the conditions in an experiment; they often feel commitment towards the study and hope that it shows what it was meant to show. This commitment, combined with their perception of the experimental hypothesis, affects their behaviour. Subjects actively try to find out what they are 'supposed to do', and pick up subtle cues from their surroundings including the behaviour of the experimenter and other subjects. These cues are known as the demand characteristics of the experiment.

EXPERIMENTER EFFECTS

Experimenters usually have expectations about the outcome of an experiment, and this can be a source of bias. Rosenthal* has estimated that about two-thirds of the errors made in recording data err towards supporting the hypothesis being tested. The behaviour of the experimenter can unintentionally pass on his expectations to the subject (see Demand characteristics). Experiments in the social sciences are unlike those in physics and chemistry in that the personality, age and sex of the experimenter may be important because these characteristics may cause the subject to behave in different ways.

ORDER EFFECTS

If a subject has to perform a series of actions, the order in which he performs them will have an effect on the efficiency of each of his actions. The two major order effects are:

(a) Practice, which causes increase in efficiency.

*R. Rosenthal, *Experimenter Effects in Behavioural Research* (Appleton-Century-Crofts, 1966).

(b) Fatigue, which causes a decrease in efficiency.

These effects occur when using repeated measures design and are usually controlled by counterbalancing (see below).

INDIVIDUAL DIFFERENCES — The personality and ability differences between subjects which could be confounding variables when using different subjects in each condition. Individual differences can be controlled by matching the two groups by pretesting the subjects for factors relevant to the study and allocating them to groups, so that the two groups are as similar as possible. More commonly they are controlled by randomly allocating subjects to the conditions (this only works if the groups are of a reasonable size).

Methods of control

The techniques described below which are designed to ensure that extraneous variables have a similar effect on both conditions, so that any difference between conditions can be attributed to the independent variable or to chance.

ELIMINATION — The removal of the extraneous variable. For instance, with the extraneous variable noise, ensuring that there is no noise.

CONSTANCY — Keeping the extraneous variable constant between the two conditions, for example, ensuring that there are equal levels of noise in the two conditions.

COUNTER-BALANCING — Control of variables that vary over time, such as practice or fatigue, by alternation of the conditions so that the variables have an equal effect on both conditions.

RANDOMISATION — Extraneous variables are varied unpredictably (by processes such as picking names out of a hat).

SINGLE BLIND — A range of techniques used to prevent the demand characteristics of the experiment influencing the results. The subject is kept in ignorance of the real nature of the condition he is in. A common example of a single blind occurs in studies of the effects of drugs on behaviour; rather than simply being given no medication in the control condition, the subject is given a placebo (a substance such as distilled water which has no medicinal qualities) so that he does not know when he is receiving the drug and when he is not. Ideally the experimenter should also be unaware whether the drug or placebo is being administered until after the data has been collected (double blind).

DOUBLE BLIND

A way of ensuring that the experimenter cannot influence the results of an experiment through his expectations. For example, in an experiment designed to see whether extroverts perform differently to introverts in a reaction time experiment, the experimenter is kept in ignorance of the subject's personality test score whilst testing reaction time.

Experimental design

This refers to the distribution of subjects between conditions. There are three main types; repeated measures, independent subjects and matched pairs.

REPEATED MEASURES DESIGN

All subjects perform in both experimental and control conditions.

Advantage: It controls for individual differences by the method of constancy.

Disadvantage: It cannot be used in situations where the experiment involves learning something under two different conditions if the skill to be learnt is the same under both conditions. For example, it could not be used for testing different methods of learning to ride a bicycle.

INDEPENDENT SUBJECTS DESIGN

Some subjects perform in condition A and others in condition B. To control for individual differences the groups should be large and randomly allocated.

Advantage: It can be used when performing one condition would affect a subject's performance in the other condition.

Disadvantage: More subjects are needed than for repeated measures and steps have to be taken to ensure that the two groups of subjects are similar, for instance by randomly allocating the subjects. The design should not be used with small groups of subjects.

MATCHED PAIRS DESIGN

From the results of a pretest the subjects are sorted into matched pairs (pairs of equal abilities on the task to be measured). One from each pair performs in the experimental condition and one in the control.

Advantage: Controls for individual differences by the method of constancy and can be used in situations where repeated measures are not possible.

Disadvantage: The choice of characteristics to match is a subjective decision. Pretesting is very time-consuming and if it is too similar to the experimental tasks can affect behaviour in the real experiment.

Population

The total number of people, objects or events from which a sample is drawn, for example, the voting population of Britain, the population of students at a college, or the population of cards in a pack.

Samples

The part of a population used in a study.

REPRESENTATIVE SAMPLE

A sample which, apart from its size, has the same characteristics, in the same proportions, as the population from which it was drawn. Having a representative sample allows us to generalise our results to the whole population.

RANDOM SAMPLE — This is a sample drawn in such a way that every member of the population has an equal chance of being selected, for instance, the use of a pin or drawing names out of a hat. This should produce a representative sample.

QUOTA SAMPLE — The population is analysed by picking out those characteristics which are considered important. Individuals are then systematically chosen so that the sample has these same characteristics. The system will produce a representative sample as long as the right characteristics were chosen in the first place.

Types of data

QUALITATIVE DATA — Data that does not refer to quantity, such as 'this room is *green*', 'subjects felt *happy*'.

QUANTITATIVE DATA — Numerical data. There are four major types: nominal, ordinal, interval and ratio.

NOMINAL DATA (sometimes called frequency data) — Refers to frequency of occurrence of particular categories. For example, there are 3 boys and 6 girls in this room. Nominal data also includes the use of numbers as mere labels, such as on footballers' shirts.

ORDINAL DATA (sometimes called ranked data) — Gives information about relative position: first, second, third, etc. For example, Red Rum 1, Blue Lady 2, Arkle 3.

INTERVAL AND RATIO DATA — Gives information about position plus relative differences. The units of public measurement are all ratio or interval data such as feet, inches, yards, pounds, ounces, centimetres, grammes, kilogrammes, degrees. For example, if we are told that Jim is 6 feet tall and his daughter is 3 feet tall, we know not only that Jim is taller than his daughter but also how much taller he is.

Interval and ratio data differ with respect to their zero points. Interval data, such as temperature measured in degrees Celsius or Fahrenheit, has an arbitrary zero point (0° does *not* mean no temperature). Most public units take the form of ratio data since they have a logical zero point which really does mean no weight, no height, no time, etc.

Hypotheses

These are predictions made before conducting an experiment about the possible outcomes. The actual experiment tests these predictions and statistical tests enable us to evaluate them.

EXPERIMENTAL HYPOTHESIS

The prediction made before an experiment that the independent variable will have an effect on the dependent variable. There are two types, two-tailed and one-tailed hypotheses.

TWO-TAILED EXPERIMENTAL HYPOTHESIS

The prediction that there will simply be a significant difference between the results of the two conditions. (It does not predict the direction of the difference.)

ONE-TAILED EXPERIMENTAL HYPOTHESIS

The prediction that condition A results will be significantly bigger (or faster, etc.) than condition B results. This hypothesis is more precise than a two-tailed hypothesis since it states the direction of the expected difference. In most cases we state a two-tailed hypothesis since we must have a very good reason for stating a one-tailed.

NULL HYPOTHESIS

The opposite of the experimental hypothesis – in other words, the hypothesis that there will be no difference between the results of the two conditions. After an experiment, the statistical test tells us the probability that the null hypothesis is true – if this probability is less than 5% we reject the null hypothesis and accept the experimental hypothesis.

Probability A measure of how likely or unlikely some event is or how likely one explanation is compared with another. It is expressed either as a percentage or on a scale from 0 to 1 (see Table 29).

Table 29

	Very likely	*Possible*	*Very unlikely*		
Percentage scale	99%	50%	5%	1%	0.1%
0 to 1 scale	0.99	0.5	0.05	0.01	0.001

SIGNIFICANT DIFFERENCE The difference between the means of two conditions is said to be significant if it is unlikely that it occurred by chance. If a result is significant it means that the same experiment repeated under the same conditions will *probably* lead to the same result (that is, if condition A was fastest this time, it will probably be fastest next time). By convention, the level of probability that the result occurred due to chance below which we say that a result is significant, is 5%. If a statistical test tells us that the probability that the result occurred due to chance is over 5%, then the result is said to be *not significant* and may be different if the experiment is repeated.

There are two types of error possible with the use of the term significant: Type 1 and Type 2 errors.

TYPE 1 ERROR When the result is said to be significant but later found to be unreliable; that is the experimental hypothesis was incorrectly accepted.

TYPE 2 ERROR When the result is said to be not significant but is later found to be consistently repeatable, that is it was a genuine result. (This often happens when only a few subjects are used in an experiment.) The null hypothesis was incorrectly accepted.

SIGNIFICANT CORRELATION — A correlation co-efficient which is probably not a result of chance factors.

Statistical tests of significance

Techniques for calculating the probability that the difference in the results of your experiment were due to chance, that is, the probability that the null hypothesis was true. There are two types: parametric and non-parametric tests.

PARAMETRIC TESTS — An example is the related t test. Parametric tests use interval or ratio data. They assume that the data in the two conditions is drawn from normally distributed populations of scores with equal variance (a measure of the scatter of scores around the mean, similar to standard deviation). These tests are more powerful than non-parametric tests (that is, they are more likely to find significance if it is there).

NON-PARAMETRIC TESTS — Examples are the Mann-Whitney, Wilcoxon, chi squared and Spearman's rho tests. These do not make the assumptions of the parametric tests. They are less powerful than parametric tests. They are easy to recognise since they use *nominal* and *ordinal* data.

p — The end result of a statistical test is a p score. This is the probability that the result occurred due to chance factors. If p = 5% or less, the result is said to be significant. (As you have probably noticed, p is equal to the probability that the null hypothesis is true.)

YATES CORRECTION — The formula amendment used for a chi squared test when the degree of freedom is 1.

DEGREES OF FREEDOM — When using statistical tables, after working out a test we usually need two values: the first is the value of the statistic (t or chi, etc.)

Table 30

TESTS TO FIND SIGNIFICANT DIFFERENCES

DATA	DESIGN Independent subjects	Repeated measures or matched pairs
NOMINAL	Chi squared*	Sign test*
ORDINAL	Mann-Whitney*	Wilcoxon*
RATIO/ INTERVAL	Unrelated t**	Related t**

*Non-parametric tests
**Parametric tests

and the second is either the number of subjects (in the case of the Mann-Whitney and Wilcoxon tests) or the degrees of freedom (for instance for t tests and chi squared tests). The precise formula for degrees of freedom (df) varies depending on the statistical test, but it always relates to the number of scores or categories that are theoretically free to vary given the total score actually obtained in the study. Imagine that you have £100 to give away and you decide to give it to your three best friends. Although there are three people, there are only two degrees of freedom because once you have decided to give, say, £50 to one friend and £35 to another, the value for the third friend is fixed (£15). The instructions given for each test specify how to work out df when it is required.

ROBUST TEST A test which can withstand some violations of its basic assumptions and still avoid errors in the interpretation of the result (for instance the related t test can perform well even when the assumption of normality made by parametric tests is not met).

POWER OF A TEST — A powerful test is one that will detect a significant difference (if one exists) using a smaller sample size than a less powerful test. On the whole, parametric tests are more powerful than non-parametric tests.

CORRELATION CO-EFFICIENT — A measure of the degree of relationship between two variables. It can be positive (where if one variable is high so is the other), or negative (where if one variable is high the other is low), and varies between +1 and −1 (0 means no relationship).

NORMAL DISTRIBUTION — A frequency distribution where the mean, median and mode have the same value. The graph is bell-shaped and symmetrical on either side of the mean. 68% of all scores lie between +1 and −1 standard deviation from the mean. See notes on standard deviation and normal distribution.

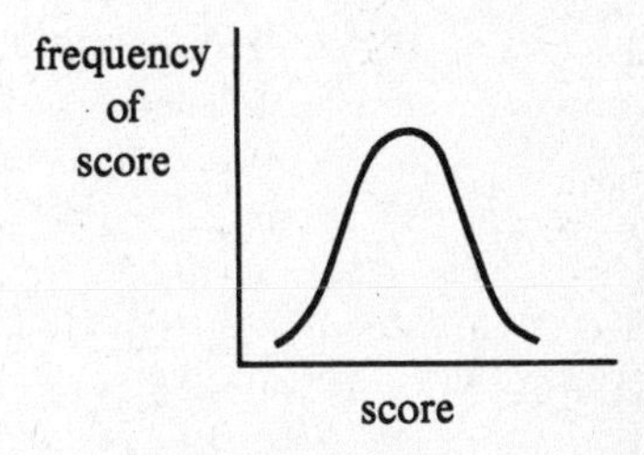

Figure 20

SKEWED DISTRIBUTION — Frequency distribution where mean, median and mode have different values. Skewed distributions take two forms: the median may have a higher or lower value than the mode.

STANDARD SCORES (Z scores) — We often need to compare an individual's score on different tests but we can't simply compare raw scores since tests differ in their difficulty, structure and scoring. Z

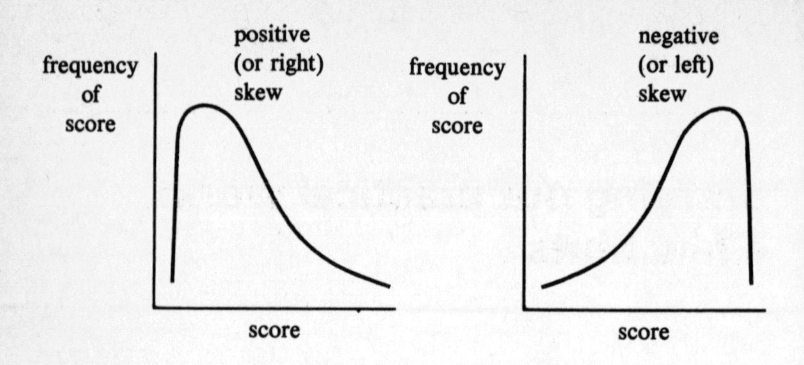

Figure 21

scores are standard scores which state how far from the mean an individual score falls. Z scores may only be used when the scores are normally distributed. Once standard scores are quoted direct comparisons can be made between performances on different tests. (See page 17).

Carrying out practical work: a few hints

The importance of planning and thorough preparation before carrying out practical work cannot be overstressed. Careful design and control are crucial. The authors have come across numerous cases of students who have had good ideas and rushed out to investigate them only to find when they came to analyse the data that it was meaningless because of some flaw in the design. Get the methodology right and the analysis follows, get the methodology wrong and you are in trouble!

First decide which *population* you want your results to apply to; you might want your results to apply to all humans (good luck!) or just to males or females, to students or to children under the age of seven, etc. Once you have decided this you will have to select a *representative sample*, i.e. a sample which has the same characteristics as the population. Remember that if you take an unrepresentative sample of the population, you might not be able to generalise your results.

Make sure any questions are clearly phrased, unambiguous and encourage the subjects to answer truthfully. Closed-ended questions which simply require a choice of predetermined answers such as Yes or No are easy to record and quantify, but may not give as much information as more open-ended questions which allow subjects to expand on their answers. The interpretation of answers to open-ended questions may be difficult and can be affected by experimenter bias. If you decide to use Yes/No or other multiple-choice-type questions, it might be useful to do a preliminary investigation using open-ended questions to help you to decide how to phrase the closed-ended questions and which possible responses should be included in the multiple choices.

If you are going to ask lots of questions, try to avoid building in a response bias which might cause incorrect answers from a careless subject. Response biases can occur when a series of questions have all been answered in the same way. For instance, a series of questions such as 'Have you suffered from smallpox?' and 'Have you ever mugged an old lady?' would all be answered 'No' by most people; if this series is immediately followed by a question such as 'Do you smoke cigarettes?' many smokers, especially those in a rush to finish, will answer 'No' to this question as well. Careful preparation of the list of questions can avoid the build-up of these habitual responses.

Check the reliability and validity of your methods of measuring the dependent variable (see page 78).

Write down your instructions to subjects and make sure that you give the instructions in the same way to each subject.

Make sure you know exactly how you are to record the subjects' responses. If there is more than one experimenter, check that you all agree on the methods of recording results.

Decide in advance how the results will be analysed. It's often too late to make major changes in the data once it has been collected.

Always perform a pilot study. This means doing the practical with a small group of people before going ahead with the full-scale study. The pilot study often shows up weaknesses in things like instructions to subjects or difficulties in measurement, and these should be sorted out before continuing with the real thing.

Take the exercise seriously; a flippant attitude on the part of the researcher will produce flippant responses from the subjects.

Interpreting results

(1) Once you have carried out the appropriate statistical tests on your data, you must decide whether to accept or reject your hypothesis or hypotheses. You have now in effect obtained the finding from your study.

(2) You must now decide how your findings relate to other comparable psychological studies. If you are performing a replication of a previous study, the comparison of findings is relatively straightforward. Even most non-replication studies are highly likely to have similarities with other studies.

(3) If your results are substantially different from other studies you will need to offer explanations for this difference. Consider the following:

(a) Was your subject sample biased/unrepresentative/too small?
(b) Were your instructions standardised? Were they the same or similar to those used by previous researchers?
(c) Did you use different measurements of the dependent variable?
(d) Did you use the same type of experimental design?
(e) Be honest! Were there flaws in the design or in the way the study was carried out?

If none of the above checklist gives you an idea for the discrepancy between your results and other published work, *always consider the possibility that you may be right and they might be wrong!*

The published study might be suffering from a Type 1 error: since most studies use the 5% level of significance there are bound to be a number of chance results that are published as significant. These studies often remain unchallenged for a considerable time because people getting different results assume that they have made a mistake themselves or that they have done the experiment slightly differently. (One of the reasons for being precise in the write-up of your method is to help others replicate your research.)

Writing up a Report

There is no single correct way of writing up a practical report, but examining boards, academic departments and your lecturers will have clear expectations about certain essential features which should be incorporated. The best first action for you to take is to look at a psychology journal (such as *The British Journal of Psychology*) and see how it is structured. Although there is a reasonable degree of flexibility about the exact format of the report, the importance of how it is written cannot be overstated. This view can be justified either in terms of psychology or of examination success: psychology, like any other scientific enterprise, depends for its development upon the communication of ideas and the research report is the main way in which we communicate our findings to other researchers. The ability to write accurate and high-quality reports is a very important skill. The following guidelines on lay-out and content should always be used in conjunction with advice given by your lecturer. Do, however, bear in mind the following yardstick: *your report should be sufficiently detailed to permit someone else to replicate your study without them having to contact you to ask questions about it.*

Title

Although the title should be concise, it should also indicate the essential nature of the study. For example, 'A memory experiment' would be both too short and insufficiently precise. 'Our fourth memory experiment' may be correct but still gives us insufficient information. The appropriate detailed title would be 'A repeated measures experimental study of recall and recognition as a measure of short-term memory'.

Abstract

This should be a short synopsis or thumbnail sketch of the practical, and should only rarely exceed one average paragraph in length. It should contain enough detail to let the reader know what you did, with whom and to what effect. It lets him know if your study is of interest to him and whether it is worth his while reading on. State briefly who were used as subjects, how the independent and dependent variables were measured, what the experimental task was, what the outcome of the study was and what conclusion may be drawn from the study. Do remember to keep this section concise; this is not the place for elaboration. The abstract should help a reader decide whether he wants to read the detail contained in the rest of the report.

Introduction

State what the aim of the study was and what hypothesis or hypotheses were being tested. Then put the study into context – as we have said previously, your study is not being carried out in a vacuum, so describe other relevant psychological research and theory. This will constitute the background to your study. As a general strategy you may find it useful to begin by describing the general area of psychology in which your study is located, then to provide a brief review of relevant studies. After you have done this, give a more detailed and critical description of perhaps three or four of the most important/relevant ones. Finally you should explain how your study differs from the ones described and how and what it adds to our understanding of this area of psychology.

Method

In this section you need to explain *exactly* how your study was carried out. This is the part between the planning and the analysis of results which in many people's minds *is* the practical. The exact format of this section may vary quite substantially according to the nature of your study (a laboratory experiment may, for example, require a different type of write-up of method to a case study), but it is customary for the section to be sub-divided. The following categories are those most frequently used:

Design
Subjects
Apparatus
Procedure

Design

This refers to the 'technical' specifications of the study. What is its methodology? Is it a laboratory experiment, a field experiment, a survey, an observational study, or what? If subjects were allocated to conditions, how was this done: repeated measures, matched subjects, independent subjects? If there are independent and dependent variables, what are they? How many conditions were used? How many trials did subjects perform? What control procedures were used? Do not write at length here about exactly how the study was carried out – such detail belongs in the Procedure sub-section. This section should be concerned only with the overall 'shape' of the study.

Subjects

The philosopher Kant argued that psychology can never be a proper science because of its subject matter – living organisms, usually human. We may choose to disagree with his conclusion, but the fact remains that the psychologist has to deal with special problems which do not confront people such as the chemist who deals with inanimate objects. It is probably fair to say that the results obtained from any psychological study are governed by the type of subjects being studied; other than the most basic studies dealing with the few 'universals' which are common to all human beings, we are faced with the fact that had a study been carried out with different subjects, a different outcome may have ensued. Therefore, it is most important that you state exactly who your subjects were, from what population they were drawn and how they were selected. Be sure to concentrate, however, only on their *relevant* characteristics. It is unnecessary, for example, to list subjects' names and what they were wearing, unless these are an integral part of the study. Consider what characteristics of the subjects may have influenced the results. Age, sex, intelligence, motivation and naivety/previous experience are usually relevant, although factors will vary according to the specific study. For example, in an experimental study of eye-hand co-ordination, subjects' handedness (whether they are left- or right-handed) would be of relevance.

Apparatus

Give a full list of the specialist apparatus that you have used, and, if necessary, explain how it was used. If the apparatus used was unusual or complex a diagram is often helpful. Be sure to include only technical or specialist apparatus. For example, it would be appropriate to include a reaction timer, but not a table or a pencil.

Procedure

This sub-section may be considered to be of particular importance since it gives you the opportunity to tell the interested reader *exactly* what you did and what the subjects were required to do. Thus, it should constitute a clear statement of how your study was carried out, and you should say how any difficulties were dealt with. Reference should be made to all instructions that were given to subjects and *all* matters of procedure should be detailed. Tell the reader exactly what happened and provide him with sufficient information that he should be able to carry out a precise replication of your study.

Results

A *summary* of your results should be presented in this section (e.g. totals, means, standard deviations). Raw data (e.g. subjects' individual results) should not be included – these should be given in the appendices at the end of the report so that the reader may inspect them if he wishes to. Present the results as clearly and concisely as you can so that the reader is able to easily appreciate them. It is up to you to do the hard work for him at this stage. Visual display (e.g. summary tables, graphs, histograms) should be used whenever appropriate and should always be clearly and exactly identified and labelled. The overriding concern here should be clarity.

Treatment of results

In this section you should present and justify the use of any analyses (e.g. significance tests) that you have applied to your data, but do not include statistical calculations, which should be presented in the appendices. State why a particular statistic was used and, *briefly*, what it tells us. This section should be particularly exact and precise. You might, for example, state what the obtained statistical value was, what the degrees of freedom or n size was, what level of significance was employed, whether one- or two-tailed tables were used, and what the consequential probability value was. Do not attempt to interpret your results at any length – this should be carried out in the next section – but make clear exactly what the result was by stating whether the null or experimental hypothesis was accepted (write down the relevant hypothesis so that your reader does not need to search back to the introduction to check what it was).

Discussion

In this section you are required to examine and critically evaluate the findings of your study. You should say how your findings relate to the hypothesis/hypotheses you put forward at the beginning and discuss the relationship of your findings to generally accepted psychological theory and results from other relevant empirical studies. In this way you should evaluate your findings in the light of existing psychological knowledge. If your hypotheses have to be unexpectedly rejected, or if your results are in clear disagreement with previous research and/or theory, you should make a comprehensive attempt to explain the discrepancy. For example, were there methodological flaws in your study? If there were, discuss them, offer constructive improvements and suggest ideas for further study which might clarify the issue. Be careful, however, to be sensible and realistic. Psychology experiments are rarely, if ever, carried out under perfect conditions, so don't write paragraphs telling the reader that there were slight fluctuations in the background noise levels, that the sun flitted in and out behind clouds on several occasions, that *The Sun* was not published that morning, etc., unless these factors are really relevant to the outcome of the study. In other words, you should be able to demonstrate an awareness of the limitations and weaknesses of your study and be able to suggest logical modifications and improvements. This section should therefore be viewed as a *considered critique* of your study, and will probably constitute the *pièce de résistance* of the whole report.

Conclusion

This should be a brief synopsis of the outcome of the study. You should offer a verbal interpretation of your main statistical findings.

References

In the main body of your report you should refer to books or papers by identifying the (main) author(s) and the year of publication, e.g. Bloggs *et al.* (1981). In the References section you should list alphabetically (by author) all of the works you have referred to, giving full details of each one. Thus for the example given, it would be:

> Bloggs, C. B., Jones, M., Smith, S., Smithers, A. and Smithers, P., *Starting Psychology*, London: Weidenfeld & Nicolson, 1981.

Appendices

It is customary that materials used in studies (e.g. standardised instruction sheets, progress sheets, raw data, statistical calculations) should be included in reports so that they are available for both scrutiny and re-usage by others. However, it is inappropriate to include such reference items in the main body of the report and consequently they are given as appendices at the end of each report. Include them as numbered sheets and refer to them whenever necessary in the main text.

When writing up your study in a format such as the one outlined above, try to bear in mind the aim of the exercise, namely, that of communicating your ideas to others. As we have said earlier, science is a social enterprise. So don't aim to be flash, just be accurate, honest and helpful.

Appendix 1 Notes on ranking

Take the raw scores from your study and use a spare sheet of paper for working out your ranks. First put the scores in ascending order. For example,

2 4 5 6 8 9 9

Give the lowest score rank 1, the next lowest rank 2, etc. for example,

raw score	2	4	5	6	8	9	9
rank score	1	2	3	4	5	?	?

The two 9s cause a slight problem but this is easily solved by sharing the ranks 6 and 7 between them, giving them each a rank of 6.5 (6+7 divided by 2). So our example becomes,

raw score	2	4	5	6	8	9	9
rank score	1	2	3	4	5	6.5	6.5

If there was another score higher than 9 it would get rank 8 because ranks 6 and 7 have already been allocated.

Once you have worked out the rank for each raw score, go back to your table of results and award each subject the rank that applies to his raw score.

Sometimes you get a lot of equal scores. The same rules apply about sharing ranks. (The ranks given to shared scores are known as tied ranks.) For example,

raw score	3	6	6	6	6	6	8
rank score	1	?	?	?	?	?	7

The 6s share ranks 2, 3, 4, 5, 6, i.e. each one gets rank 4 (2+3+4+5+6 divided by 5).

Rank scores are always whole numbers or a whole number plus 0.5.

When you get a lot of tied ranks it becomes tedious to add up all the shared ranks and divide by the number of shared scores. An easier way of doing this is to realise that the tied rank to be awarded is simply the middle one of the shared ranks. For example, if you have 7 raw scores of equal value which share the following ranks, each score is given rank 9.

shared ranks 6 7 8 9 10 11 12

If there are an equal number of shared ranks there is no one score in the middle and so the tied rank awarded is halfway between the two middle scores. For instance, if the following are the shared ranks, the tied rank awarded is 6.5.

shared ranks 4 5 6 7 8 9

Appendix 2 A 'master' pie chart

The pie chart below has been divided into four segments each of which is 25% of the whole. One of these segments has been further subdivided into 1% and 10% segments. You can use this 'master' to present your own data in pie-chart form by tracing the appropriate lines from the master.

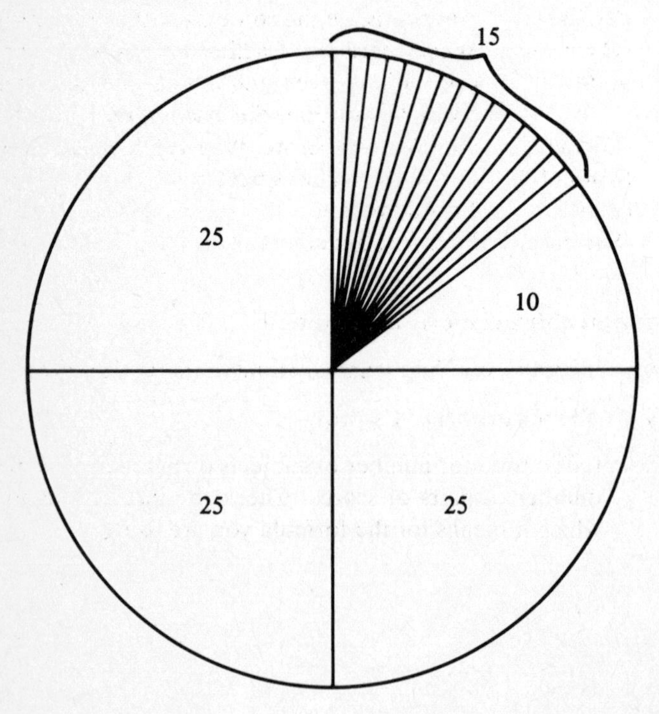

Appendix 3 Some simple rules for working out formulae

When faced with formulae such as those for the t tests, some people have difficulty deciding which bit to tackle first. The BODMAS rule can help with this problem because it helps you remember the order in which calculations must be done.

B	(Brackets)	Always tackle the contents of the brackets first.
O	(Over and under)	When there is a fraction always work out the calculations over and under the line before trying to work out the fraction itself.
D	(Divide)	Do divisions before other types of calculation.
M	(Multiply)	Do multiplications next.
A	(Add)	Now additions.
S	(Subtract)	Finally subtractions.

Common abbreviations in formulae

Σ 'Sum of', e.g. Σx means the total of all the X scores.

$\bar{X}$ The mean of the X scores.

N Often used for number of subjects or number of scores or number of pairs of scores. Check the instructions to see which it means for the formula you are using.

Index